The Heavenly Horses

by
Virginia Weisel Johnson

in co-operation with
Tex Johnson,
Animal Health Technician and
Hungarian horse trainer

MOUNTAIN PRESS PUBLISHING COMPANY
MISSOULA, 1986

Library of Congress Cataloging-in-Publication Data

Johnson, Virginia Weisel.
The heavenly horses.

1. Hungarian horse — history. I. Johnson, Tex.
II. Title.
SF293.H79J64 1986 636.1 86-21776
ISBN 0-87842-207-2
ISBN 87842-208-0 (pbk.)

Acknowledgements

Ernst and Lil Peterson, Hamilton, Montana
Bud and Sal Ruark, Hamilton, Montana
Charles Casselman, Hamilton, Montana
Sharlene Ashcraft, Hamilton, Montana
Jim and Helen Edwards, Ennis, Montana
Wanda and Steve Cooksley, Anselmo, Nebraska
Judith Gyurky, Virginia
Margit Bessenyey, New York
Francis Bessenyey, New York
Eva Bessenyey, New York
Henry Hyde, New York
George Sacellary, Batesville, Virginia
June Gilchrist, Missoula, Montana
Peter and Gene Samulevich, Hamilton, Montana
Ernst Szechenyi, Florida
John J. Slonaker, Chief, Historical Reference Branch, United States Military History Institute, Carlisle Barracks, Pennsylvania
Dr. Joseph C. Hovarth, Missoula, Montana
Dr. J. Veasekloy, Budapest, Hungary
Willie Couture, Hamilton, Montana
Kevin Grieves, Missoula, Montana

Table of Contents

Mare and colt. Ernst Peterson

In memory of Margit Sigray Bessenyey and Judith Gyurky who established the breed of Hungarian Horses in America.

Hungarian horses. Ernst Peterson

I
Countess Margit Bessenyey and her Horses

I will never forget that day in early autumn when the cottonwoods were turning yellow. Margit Bessenyey and I were looking at a bay gelding that was standing at the far side of the paddock. He was typical of the Hungarians — fifteen two, with a broad chest, head well set on his shoulders, a short back and straight legs with flat bone. My impression was of an athletic, intelligent animal. Most of the horses were friendly, but this horse would not come near us. He was wary and distrustful.

Margit said, "When a horse does not suit a buyer, I take the horse back and find another buyer."

This seemed good business to me. Such a policy could earn Margit a reputation for honesty that would attract customers who had learned to distrust most horse dealers.

"Couldn't the woman who bought this horse ride him?" I asked.

"No. She overestimated herself and I think she yelled

at the horse and let him know she was afraid. Then she hired a man who used a whip and spurs on him. The horse threw him. You can't do that with a Hungarian; particularly one with Furioso breeding."

Furioso was a bloodline in the Hungarian horses, predominately Thoroughbred. Little by little I was learning about the Hungarian horses — that they were not for sloppy, inexperienced riders or for owners who tried to discipline them by force. Handled correctly, they were loyal, obedient, responsive and never gave up. In horsemen's terms, they had heart.

Margit and I shared a common bond with the horses. My grandfather and Margit's, who had been Marcus Daly, one of Montana's Copper Kings, had been business associates and friends and now, fifty years later, Margit and I discovered mutual interests. Margit was a knowledgable horsewoman — intelligent, well-read and a stimulating companion. As a girl, she had lived on a great estate in Hungary and knew how horses had been bred there. Her mother, a daughter of Marcus Daly, had married a Hungarian Count.

For centuries, Warmbloods had been well-known, but it was only recently that they had been imported to the United States, where they soon excelled in dressage and Grand Prix jumping; however, one breed that was not promoted and that was, therefore, overlooked was the Hungarian. This was partly due to Hungary being behind the Iron Curtain and the home breeding establishments having been destroyed by the Russians during World War II.

Margit and Judith Gurky, a Hungarian who had a farm in Virginia, were determined that they would establish the Hungarian horses in the United States and make them as well-known as the Swedish Warmbloods, the Hanoverians and the Trakeners. Judith Gurky, born a Countess, and Margit, also born a

Countess, felt they had the knowledge to make a success of their project and make money too.

What Judith and Margit did not realize was that they were products of an age that had vanished in the War. They were unprepared for the commercialization of the horse world where stallions were syndicated for millions of dollars, where shares were bought and sold in race and show horses and a champion Grand Prix jumper could sell for seventy-five thousand dollars. People who owned shares in the horses never saw the animals nor cared to see them. All they wanted was a return on their investment. The operation of the stud farms was left to high-priced trainers and professional riders.

Judith and Margit, on the other hand, believed the owner should take an active part in the operation of the stud farm. Judith had little help — a couple of girls and an old colored man and, at intervals, her son, who was a Merchant Mariner who went to sea to earn money to help Judith. George was a good horseman and, in some respects, was more practical than his mother.

Margit's ranch, which had been established by her grandfather for his race horses, still retained the Victorian atmosphere of the Nineties. A large brick stable with a cupola on top stood on a green lawn shaded by cottonwoods. Inside, the stable had tie stalls and two box stalls, where Margit kept her two Hungarians — a mare named Brownie and a gelding called Taszilo— both of whom had been trained in the East. A lane overhung with cottonwoods led past the Manager's white clapboard house, the bunkhouse and a row of red barns and corrals. The corrals faced a mile and a quarter race track. At one end of the track were sheds where the vans were stored that Margit used to transport horses from Montana to the Cooksley Ranch in Nebraska, to Judith's in Virginia and to Margit's Maryland

Hungarian mares at the Stock Farm. Ernst Peterson

property for showing and sales in the East. The Cooksley Ranch was the third Hungarian stud farm in the United States and the Cooksleys, Margit and Judith worked closely together on stallions and bloodlines. As yet, there were so few Hungarians in America that care had to be taken to prevent inbreeding.

Of the three stud farms, Judith's was ideally located in Virginia, which was hunt country and where steeple-chasing was popular. Most people rode Irish hunters or Thoroughbreds. In an attempt to inform people about the Hungarians and, incidentally, to make a little money, Judith wrote "The Mark of Clover" in which she related the legends of the Hungarian horses. Judith published the book herself, but while extracts were printed in various equine magazines, Judith's stories did not make much impression on the American horse world where breeds such as the Morgan, Saddlebred, Thoroughbred and Arab were being efficiently publicized. Another breed, the Lippizaner, was also introduced romantically to Americans by the Disney film "The White Stallions."

Judith's stories remained unread by the public and, what was worse, brought no financial return. Judith badly needed money to feed her horses. Indeed, she always needed money for her horses. Judith was also bothered by what the American public wanted in the way of sport horses. At the moment, Thoroughbreds off the track appeared popular. Indeed, Judith told me that a buyer had taken two Hungarians and resold them with false papers as Thoroughbreds. The Hungarians, Judith was afraid, were not tall enough. They averaged fifteen hands to fifteen three, while horse show jumpers were usually sixteen to seventeen hands. Should they be trained as event horses or for dressage? Or, as Margit wanted, all-around horses that could perform in competitive Rides, as well as driving and jumping.

The Hungarian had always been an all-around horse, but was that profitable today? John Brewster Darley had bred for speed when he stood his stallion at stud and the Darley Arabian became a founder of the English Thoroughbred. In Arabia, the owners bred the Hejd racers for speed and endurance, enabling them to ride into battle on a horse "swift as the wind and tireless as the wolf."

Meanwhile, Judith continued to enter her horses in the local Charlottesville shows, while Margit concentrated on long distance Rides to prove the endurance of the Hungarian breed.

Judith and Margit were to keep this age-old European relationship with their horses. Their personal decisions resulted in the Hungarian Warmblood as he was when he came to the United States and as he is today.

II
From the Steppes of Asia

Most of the history of the Hungarians I learned from a Budapest scholar whom Margit had invited to the ranch — a tall, thin man, humorless and bleak. Shortly after his visit to the United States, he died, which was a loss to historians everywhere. His name was Miklos Jankovich and he had recently published a book entitled "They Rode into Europe," a history of Hungary with many interesting items about the horses that played so important a part in the turbulent past of Eastern Europe.

The only chapter that I felt could be improved was the one on horses in Western America, so we took Mr. Jankovich to a local calf-roping contest to give him first-hand knowledge of cow ponies, Quarter Horses and stock saddles, but Mr. Jankovich was plainly bored and after a short while we left. Jankovich, I felt, thought that American horse events were new and crude. He was happier to return to the days of the

Turkish invasions and Attila the Hun. Horses, to Jankovich, were not living animals that we rode and were fond of; they were means of locomotion for the hordes that overran Hungary, as impersonal as an automobile.

Nevertheless, "They Rode into Europe" is a well-documented and fascinating book, which, I'm afraid, did not sell well. I reviewed it for "The Chronicle of the Horse" and was thanked by Mr. Jankovich, but the man was not comfortable in America. Did he realize that he had been passed by in a world that was changing rapidly?

Of Judith, Jankovich would not speak. Her book, "The Clover Horses," according to Jankovich, was beneath notice, a mixture of fancy and romantic inaccuracies. Jankovich researched, documented and referred to other historians, while Judith relied on tales told by horsemen around campfires and in the stables of old coachmen. In addition, records were lost or destroyed during frequent wars. An American Lieutenant of Remount in World War II told me he came on a small herd of Hungarian horses in the care of a mustachioed groom who begged him to save horses and stud books from the Russians.

"I have no place for horses or stud books," the Lieutenant said, and drove on in his Jeep.

Fortunately, some of the horses were saved by the Americans and their stud books, too, while owners of big estates and government stud farms managed to rescue important records. Still, the fact remains that few countries experienced the turmoil of war and pillage as often as did Hungary, which lay on the invasion route of the Huns, the Moslem Turks and the Russians. Each invasion and occupation left its mark on the Hungarian horse, such as the tough little ponies of the Huns and the Turkish Barbs.

So what was the Hungarian horse? What was he

when Margit and Judith and the Cooksleys established their stud farms?

Jankovich and later historians wrote that long before the Christian era Nomadic tribes of the Asiatic Steppes caught and trained native ponies. They also hunted them for meat and drank the milk of the mares. The horses were very small and coarse and were as often packed as ridden.

According to Miklos Jankovich, there were three original races of the true Eurasian horse — the pony that lived on the edge of the forests of Western Europe, the Taki that once grazed on the borders of Soviet Asia and Mongolia and that exists today only in zoos where it is known by its Latin name of *Equus przekalskii poliakoff* and the Tarpan that once roamed the borders of Soviet Asia and Mongolia. The Tarpan formed the foundation stock of the domestic horse.

However, another classification of the ancestors of the domestic horse was made by H. Ebhardt, a German scientist. Ebhardt's system, based on feeding habits as well as physical characteristics, listed as follows: two pony types about twelve hands in height, stout, short-legged, with coarse heads; and two horse types, one with a Roman nose, long ears, sloping croup, usually slab-sided and standing fourteen to fifteen hands. The second horse type was short-backed with a tail set high, a dished face and, in general, a more refined animal than the number two horse. He stood eleven to twelve hands and closely resembled the Arab of today.

About the same time there was another horse — the Turanian, known as the "heavenly" horse — whose home was the northern slope of the Afghan mountains. The Turanian, also known as the Turkoman, was bred by the tribes of Turkistan. Alexander the Great was reputed to have mounted his cavalry on Turanians, while a Chinese Emperor, hearing of their beauty and

Mongol on horseback. Budapest.

speed, offered a king's ransom for a stallion. The Turanian, often a golden dun in color, had long legs, a slightly dished face and a highly developed nervous system, which made him as different from the Taki and Tarpan as a Thoroughbred from a cayuse. A horse such as this that had been killed several hundred years before Christ to accompany his owner to Paradise, was found in the Pazyryk burial mound in the Altain mountains of Asia. This area of Far Eastern Europe and Asia existed in a state of constant warfare when nomads roamed the Steppes for pasture and water and great nations fought for power. The Persians and the Greeks established empires and, when they crumbled into dust, there were the Arabs and the Seljuk Turks who slaughtered and pillaged. Always the visitors seized a booty of slaves and horses. The horses they cross-bred with their own animals, which resulted in a mingling of the Tarpan and the Turanian, and when the Asiatic invaders swept across Africa and Europe, the Oriental bloodlines became mixed with the West European horse.

In 376, Attila the Hun rode out of the East to ravage what was left of the Roman Empire, his hordes galloping over what became Hungary, Italy and France, and after the Huns came the Avars and then the Mongols led by Genghis Khan. The Mongols rode Tarpan-type horses — fast, tough little beasts. Loose herds of horses followed the Mongol armies for use as reserve mounts, for meat and for blood which the Mongols drew from the veins of their animals and then solidified into a pudding. These Mongols were superb horsemen whose weapons were a short bow and a sword. "The Secret History of the Mongols," an account that is source material for the Mongols, says that the Mongols practically lived on their horses and that, contrary to later historians, they treated their horses well. They were afraid not to, for Genghis Khan threatened death to the careless warrior. The men were not to tie bundles behind their saddles nor waste the horses' energy by chasing game.

When Genghis Khan was killed in a hunting accident, the Mongols retreated to the Steppes of Asia, but behind them they left ponies that cross-bred with native horses.

After the Mongols, the Turks invaded Eastern Europe under the green banner of Islam, burning and pillaging the cities of Hungary and the countryside. For one hundred fifty years the Hungarians fought the Turks. The land lay in ruins and the horse herds dwindled, but some native Hungarian mares and stallions mingled with the Turkish horses, known as Orientals, a small animal but finer-boned and more elegant than the Mongol-type pony.

Horses were the Hungarian's pride and, indeed, influenced his way of life. They provided transportation over muddy, rutted roads, a method of communication in a primitive country. They symbolized wealth and status. Horse skulls were hung on the timbering of living rooms in villages for protection against evil spirits and the owner of a superior animal was envied and admired. In pre-Christian times, a Hungarian was buried with his horses and horse equipment to provide him transportation into the next world.

Over the centuries, the Hungarian horse became well-known in Europe. Miklos Jankovich wrote, "The Magyars are the only pastoral people who possess written evidence of their centuries of stock breeding." In addition, the Magyars were among the earliest people to geld horses. As a result, the French word for gelding is "hongre," which means Hungary.

"The Hungarians," an Englishman told me, "felt much the same about their horses as do the Irish, except

Tschingis=Chan als Nomadenhäuptling

Ghengis Khan. Hungarian version of ***Secret History of the Mongols.*** Budapest.

the Irish are more sentimental. The Hungarians were more practical; they had to be — their horses meant their survival.

In time, the Magyars became less nomadic and their leaders settled on great estates where they raised their horses. In the Eleventh Century, their horses were so popular as cavalry mounts that the King of Hungary forbade their export, although they still resembled the ponies ridden by the hordes of Genghis Khan. In ensuing decades, a need was seen to improve the Hungarian, so in the Seventeenth Century when the Turks were expelled from Eastern Europe and Hungary became part of the Austrian Empire, the government established four stud farms and began a program of controlled breeding. The small, bad-tempered ponies might have served the Magyars, but the time had come for a larger, faster animal for the Imperial Cavalry.

James Fillis, considered the greatest European high school rider in the 1890s, wrote in *Principes de Dressage et Equitation*, "In my opinion, the Halfbred is the best animal for war." This included the Hungarian, as well as some French and German animals.

In Hungary, the government breeding program was under the Department of Agriculture, established in the Nineteenth Century. During this time, the government established a registry for horses approved by the Department of Agriculture, which was strictly regulated until World War II when the Communists took over the disorganized remains of what had been a thriving equine industry.

Margit could not register her horses with the People's Republic, so she and Judith, as well as the Cooksleys, established their own registry, much as had the big estate owners in the past. Judith had papers on the horses she had brought from Hungary. Margit had papers from the Remount on her mares and stallion and the Cooksleys had papers on the Remount stallion they were breeding mares to, so the basis for the Hungarian breed in America was legal.

The foundation sire of Margit's horses was Honpolgar IV, conceived in Hungary, born in Germany where his mother foaled after being stolen from her native land and then "liberated" by the American Remount Service from the Nazis for shipment to the United States.

A registered Thoroughbred mare or a purebred Arab could be bred to Honpolgar or any Hungarian stud and be registered as a Hungarian. This had been the case ever since the registry was established and is similar to breeding practices in other European countries.

Margit's stud books were kept by a middle-aged Hungarian woman whose name was Hannah. During the Fifties, Hannah had been imprisoned by the Communists and had only attained freedom when Margit had helped her escape. During the winter Hannah worked as a translator in Washington and for a few weeks in the summer she checked the stud books.

"I need a Hungarian," Margit said, "because the horses have Hungarian names."

Each foal born in the Spring was given a page in the stud book with a photograph and a diagram of who the sire was, who the dam was and then who the dams and sires were of the mother and father. By reading the diagrams it was possible to tell what bloodlines were predominant in the animal — Furioso or Gidran or Shagya or, in some instances, Judith's Clover horses. The ancestry of some of the colts could be traced back a hundred years.

In the Turkish wars, as well as other conflicts, the Hungarian light cavalry had played an active part. Indeed, the Hungarian Hussars, superb horsemen and fierce fighters had served as mercenaries in Europe where they and their horses gained a reputation that won them a place in history. A horse that could carry weight, endure privation and be capable of speed was necessary to survival. The Hungarians had already established a stud at Mezohegyes in 1785. Babolna was named a government stud in 1789 and Kisbir and Fogaras in the mid-Nineteenth Century.

In 1785, the first horses were purchased for Mesohegyes. The mares were Hungarians, Turkish and Holstein. The Holstein were horses from the East Prussian province of that name where the Germans had turned from the large, clumsy destriers ridden by the knights to Andulusian and Neopolitan chargers crossed with Thoroughbreds. Records do not say, but it is possible these mares had some Turkoman as well as Akhal-Tekke breeding, the last being the fabulous horses that resembled the Thoroughbreds and that came from Turania, an area on the Soviet-Afghanistan border. The German Trakener stud was founded by a stallion of this Turkoman-Akhal-Tekke blood.

In the early Eighteen Hundreds, a stallion that was to become well known among Hungarian horses came to Mezohegyes. His name was Nonius — his sire the English Thoroughbred Orion and his dam a Norman mare. A bright bay, he was foaled in 1810 and when mature stood sixteen hands three and one-quarter inches. He founded the Nonius line of Hungarians valued as coach horses for hauling artillery and as cavalry mounts. Not long after (1820), a number of selected mares was sent to the Arab stud at Babolna to be bred to the Arab stallion Gidran, a brown horse that stood fifteen hands one inch. The foal of one of these mares, Gidran II, became the foundation sire of the Gidran line.

Offspring of the Gidran and Nonius line kept the family name, while the characteristics of each breed were emphasized in the future. For instance, English Thoroughbred and French coach horse blood were bred into the Nonius and English Thoroughbred and Arab into the Gidran.

In 1838, a number of chestnut mares were selected to become the foundation stock of the Gidrans, who were all chestnut in color.

In 1841, another foundation sire stood at Mezohegyes, an English Thoroughbred named Furioso, a cherry bay sixteen hands two and one-half inches, whose sire was the English Thoroughbred Privateer and his dam the English mare Whalebone. Furioso had been raised by a Hungarian nobleman who sold him to the government stud. The owners of large estates frequently sold or leased animals to the government, in turn leasing or buying horses from the Department of Agriculture.

The bloodlines of the Furioso, Gidran and Nonius were incorporated into stud books of the Hungarian Halfbred and are recognized today. Another bloodline equally important was the Shagya developed at Babolna, a stud devoted to the breeding of purebred and halfbred Arabs. In 1836, a Hungarian Baron was sent to Aleppo and Damascus where he purchased nine Arab stallions and five Arab mares. One of the stallions was the famous Shagya. Within the next fifteen years, more mares were bought in Egypt and Syria. Only the best animals were kept for breeding. Shagya was a flea-bitten grey with cinnamon-colored freckles, standing approximately fifteen hands. Another valued stallion was Mehemet Ali, a chestnut sired by a purebred Arab stallion out of a purebred Babolna mare. Halfbreds as well as purebloods were branded with a capital B under a crown on the horse's back near the withers where it could be covered by a saddle.

The Commandant of Babolna for many years was Colonel Michael Fadlalleh el Hedad, a Bedouin Arab who came from Arabia as a poor, untutored boy escort-

Furioso stallion. Budapest.

ing one of the stallions. Hungarians today living in America whose parents knew Hadad speak of him with admiration — the Bedouin boy who grew up and was educated in Hungary and became famous for his knowledge and understanding of his beautiful horses.

In 1911, Spencer Bordon, an American who visited Babolna, described the remarkable relationship between the Colonel and the Arabs at the stud farm. In one of the stables, three- and four-year-old stallions were tied in stalls with only a single bar between them, but not one of the horses fought the other; and so it was with the mares who, in another stable, were tied in two rows with no barrier between them and yet remained quiet except when Hadad appeared. Then many whinnied for attention and a lump of sugar. In still another stable, mares were tied beside mangers, their foals free to wander about as they pleased. When they had been fed oats and hay, they were turned out on the plain to be watched by mounted horse herders. Even free in the pastures they were curious and gentle, crowding around visitors.

To Colonel Hedad, the Arab was the most desirable horse. He was fleet, sure-footed, intelligent and loyal to the death. When Spencer Bordon asked him — thinking of cavalry horses that were usually larger — which horse Hedad preferred, a small or a large horse, the Colonel replied, "The small ones every time!" In a hundred large horses, he said, he found it difficult to select ten good ones, while in an equal number of small horses it was hard to find ten poor ones.

In a herd of Hungarian Halfbreds, which might include Furioso and Gidran breeding, it was easy to spot the horses with Shagya characteristics. They were apt to be smaller than the others, fifteen hands to fifteen two, with intelligent eyes and slightly dished face, as well as the short back and naturally raised tail.

A third stud, Kisbir, became as well known as Babolna. During the Civil War with Austria, the large estate of Kisbir was confiscated and the owner, Count Batthyanyi, was shot. In 1853, Kisbir became a government stud, ideal for the purpose with nearly forty thousand acres of pasture, large fields, a park and a race track, as well as numerous barns, stables and paddocks. In 1870, the government decided to make Kisbir the center for Thoroughbred breeding and so transferred the Thoroughbreds from Mezohegyes to Kisbir, which had the mild climate suited to English-bred animals. One of the most prominent stallions was Buccaneer, purchased from Lord Dorchester, a dark bay that had won many races in England and whose son, Kisbir, won the English Derby and the Grand Prix in Paris. Kisbir lived to be thirty years old. The most successful stallions bred only Thoroughbred mares; the others, for a lower stud fee, were allowed to breed Halfbreds. Interestingly enough, there was much opposition at first to the introduction of Thoroughbreds from horsemen who favored the Asiatic bloodlines that had been the basis of the Hungarian horse. One prominent breeder, Count Stephan Szechenyi, was criticized for importing English blood stock, as well as grooms, trainers, jockeys and hunt servants. Baron Joseph Wenkheim, an experienced breeder, wrote in 1815, "The Hungarian horse is closely allied to the Arab; it comes from the same stock of a warm climate. We must, therefore, only use as foundation stock horses from such a climate or one similar to our own, but hold aloof from all other strains and try to eliminate them, for they bring us nothing but trouble; they may have a showy exterior, but they beget nothing but bastards.

"Therefore, for our purpose, the most commendable are the Turk, the Barb, the Persian, Circassian, Caucasian, Turkoman — also the Andulusian and the

Kisbir. Hungarian Thoroughbred stallion. Budapest

Neapolitan... But all Northern breeds we should eschew; they have no common ancestry with ours and they degenerate in our climate."

However, Count Szechenyi succeeded in making hunting and racing such popular sports that the Thoroughbred became a recognized factor in Hungarian horse-breeding.

The fourth stud, while not as well known as the others, was Fogaras, which was established to breed horses that adapted to the mountains. The Lippizans were chosen as the best suited for this purpose and Fogaras the most desirable site, since it was in the mountains and the Lippizans had come from the Karst hills of Austria where they had been bred for generations. The foundation of the stud was five Lippizan stallions that had been at Mezohegyes. The mares were Lippizans, Arabs and Transylvanians. The Lippizans were good-tempered and willing to work. The peasants liked these qualities in their farm animals, while the military often bought Fogaras horses for cavalry and to drive in light harness.

The Hungarians felt that environment was very important in development of the characteristics of horses — the climate, the chemical contents of the soil, the drinking water (or lack of it) and the geography. The studs at Kisbir, Babolna and Mezohegyes were only about three hundred to three hundred fifty feet in altitude and were located on or adjcent to the Alfold, similar to the Asiatic Steppes. On the Alfold could be found sand, marshes and rich soil that produced excellent hay and oats. The three stud farms mentioned raised their own feed. Records were kept of the grain and hay fed to each animal and regular inspections were made by veterinarians.

Babolna and Kisbir each had a race track where young mares were raced. The distance was one and seven-eighths miles and the horses carried weights up to one hundred twenty-six pounds. Mares who could not do the mile and seven furlongs in about three minutes fifty seconds were sent to Budapest to be sold at auction. Spencer Bordon was told at Babolna of a pure Arab, fourteen hands in height, who was never beaten in a race and who broke a record by running the mile and seven-eighths carrying full weight in three minutes thirty-eight seconds.

Mares and stallions were both sent to cavalry regiments to see if they would make hunters and records were kept of each animal's performance.

At every stud, the purpose was to breed young stallions to be leased by private breeders so that the quality of the Hungarian horse would be maintained and improved.

Other countries such as Sweden and Germany had government studs, but these countries did not rely on horses to the extent that the Hungarians did. Despite being a part of the Austro-Hungarian Empire, Hungary remained Medieval in many ways. The Turkish occupation and constant wars had discouraged a Renaissance such as occurred in Italy and France and later, when Northern Europe and the United States became industrialized, Hungary remained largely an agricultural country of large estates and peasants who tilled the crops. Even after dissolution of the Austro-Hungarian Empire and a Communist uprising that followed World War I, Hungary was little changed. Political reforms were made under a regency, but the large estates were still held by titled landowners and the government studs still bred pureblood and halfbred horses, as they had for over a century.

Judith Gyurky and Margit Sigray Bessenyey grew up on such estates where their parents raised horses, so from childhood they knew the traditions of the Hungarian horse.

Mokus I. Mark of Clover. Ernst Peterson

III
Countess Judith Gyurky's Farm in Virginia

The first time I saw Judith Gyurky at her farm in Virginia she came in from the stables wearing a pair of farmers' irrigating boots, a disreputable skirt, the tattered remnant of a plastic raincoat and a scarf tied peasant-fashion around her head. She was in her seventies and had been swamping out box stalls.

Judith received us in the dairy barn that she had converted into a home with the original concrete floor and stalls for rooms. Near the door to the stable yard was a small cast-iron stove and adjoining the cooking area was a sink and a stall that served as a pantry and storeroom. The only heat was provided by the stove around which we huddled.

Judith could not have been more hospitable. "Will you stay for lunch?" Looking around vaguely, "I think we can find something to eat."

The Hungarian friend who had driven us from Margit's place in Maryland said quickly, "No, thank you,

Judith. We ate on the way down from Margit's."

This was not true, but we did not want to make more work for Judith.

Judith was relieved. "Then I'll show you the horses."

She led us to the stables where in nearly twenty box stalls stood mares, colts and stallions. The stalls were strewn with fresh straw and the feed boxes were full of grain that the horses munched eagerly. In the paddocks around the stables and in the pastures were more horses, mostly bays and browns with a few dappled greys. The horses averaged fifteen three, had broad chests and well-muscled hindquarters. Their heads were intelligent, their ears alert.

"That is the kind of ear I like to see on a horse," Judith told us. "It means he is watching, that he can hear what is going on. The ear is very important. Always watch your horse's ears when you are riding."

Judith pointed out the horses she was going to enter in local shows in hunter classes and in jumping and talked of sales she attended at well-known Virginia auction yards. Argo, a niece, did the show riding, as did Judith's son George, who was at present a ship's officer on a freighter.

"The horses must be shown to sell. That takes time and money." Judith explained, "so George has to earn money by going to sea. Last month he was shipwrecked off Africa."

"Why are the horses hard to sell?" I asked.

Judith shrugged. "In this country people do not know of the Hungarian. All they want is a Thoroughbred. They buy a horse off the track for their twelve-year-old daughter. The horse runs away; he throws the child. It is very dangerous."

"Don't you advertise?" I asked. "I mean put ads in equestrian magazines like the Quarter Horse and the Arab people do?"

"To advertise costs money," Judith said. "Once people learn of the Hungarians, we will sell our horses. Already we are doing better than last year."

Clinton, a colored man, was Judith's devoted groom and Dollar, Judith's husband, also helped. Dollar, an inveterate smoker, was a large, homely man who seldom spoke, probably because he was partially deaf and hearing aids were expensive when the horses needed oats or veterinary bills had to be paid.

Halfway through the tour I asked Judith if she still had a Clover horse and she exclaimed, "Yes, yes, I do! Mokus II, the daughter of the first Mokus, a mare who was out of Igezo, the Clover I rode in competition in Hungary. There she is over there."

Judith slopped through mud to a small paddock where several horses were standing, dejected, beneath a dead apple tree. Mist blotted out the surrounding hills and a faint drizzle dampened our clothes. One horse stood apart, a small brown mare with short legs and a big head. The drooping ears were nearly as large as a mule's.

"Come, Mokus!" Judith called.

The mare looked at us but did not move. Judith, reaching in her pocket, found a sugar lump. "Come, Mokus!"

This time the mare deigned to walk to the fence, slowly and deliberately, stopping every now and then to show Judith she was doing her a favor by coming and that she must not be hurried. When another horse tried to join her, she nipped him.

Judith regarded the mare with pride. "She is not beautiful and she kicks and bites, but she has the Mark of Clover. There, can you see the dark stain on her muzzle? Does it not remind you of a four-leafed clover? You must know the legend; you read my book."

When we left, Judith waved a cheerful good-bye from Mokus's paddock. "Please come again."

Our Hungarian friend shook his head as he started the car. "To Judith that ugly mare is the most important horse on the farm."

It was still raining and the road that wound up ridges and through the leafless forest was muddy. From one of the ridges we could look down on Judith's farm in a narrow valley — the frost-seared pastures and the buildings, but no hay meadows and little level ground. Beneath the grey sky it looked isolated and shabby.

"Does Judith have to buy her hay and grain?" I asked.

Our Hungarian friend said she did. "She never has money, that one, but it is all the same to her so long as she can feed the horses."

★ ★ ★

On the drive north to Maryland, I reflected on Judith's devotion to the little brown mare who might have served as a model for the ponies Genghis Khan rode across the Steppes of Asia. Judith had said Mokus had Arab and Thoroughbred blood in her ancestry, which had not been apparent. Were there throwbacks after many centuries? Were the genes of the wild horse of Asia dominant? And did the distinguishing Mark of Clover on the muzzle only appear on the pony-type horses?

This was a question that Judith could not answer; however, she said that she had only seen the Clover mark on a pony-type horse and the animal was usually of what Judith called the Clover line of breeding. The Hungarians, Judith inferred, did not ask the whys or the wherefors; they accepted it as a miracle, but not a Christian miracle, for the legend of the Clovers had originated in Pagan times before the birth of Christ.

Much later, on other visits to Virginia, Judith told me more about the Clovers and her involvement with the Hungarian horses. She spoke often about the Clovers, for her horse-breeding had been influenced by the legend.

What was the legend?

According to the centuries-old story, the family who owned a horse with the Clover stain on its muzzle would prosper and the nation in which the horse lived would exist in peace, but if the Clovers disappeared, the family would be ruined and the nation destroyed.

During one of these visits, I stayed several days and Judith gave me a room in the guest house adjoining the dairy barn. The room had been Judith's mother's and she had spent her last days there, waited on by Judith. The Countess had been a famous hostess and a beauty in Budapest and Judith had brought her to America when conditions in Hungary became chaotic. She knew nothing of the harsh realities of life. She never did know, for Judith saw she had every comfort and the courtesy to which she had been entitled in Budapest. Windows looked out on a pasture where horses grazed, an antique chest stood in one corner with a valuable icon above it, an Oriental rug covered the floor and a tapestry-like quilt covered the bed. Both rug and quilt were badly worn, plainly saved from more affluent times in Hungary.

The aging Countess, the spread and the horses were part of a picture. Judith had brought Hungary with her to America, not from nostalgia or duty, but to reinforce her belief in the legends in time of need. The horses could be judged by prospective buyers, but to understand them, one had to understand the legends. To Judith, the horses were the "heavenly horses" of the ancient Turkomans, their existence woven into history with the warriors who rode them.

How could this be explained to parents buying a show horse for a child who would outgrow it? Or to a professional trainer who would hot-rod the horse to an instant jumper?

IV
Judith and the Mark of Clover

Judith had ridden a Clover as a child on her grandmother's estate in what is now Czechslovakia. Sarga, a dun-colored mare hardly more than fourteen hands, had a coarse head and short legs. She was a tough little animal as were most of the Magyar crossbreds, and, like them she had a bad temper. On her muzzle she bore the Mark of Clover. To the old coachman, she was the most valuable animal on the estate. Uncle Pista, as the coachman was called, had no use for the "spoon-faced" Arabs and "spindle-legged" Thoroughbreds that the government and some landowners were breeding to the original Hungarians; however, Uncle Pista had to admit that Sarga's grandmother had been sired by Kettledrum, an English Derby winner.

"That is so," the coachman argued, but he pointed out that Sarga's mother was a Clover descended four hundred years ago from the Clover stallion Bayram who had helped his mistress rescue her lover from the

Turks. Bayram, in turn, had been descended from Jegyes, the golden mare who had leaped the walls of a beseiged castle to aid in the rescue of her owner, the daughter of a Magyar high priest. Jegyes had kicked and bitten at the terrified enemy, who screamed, "———soul-eating devil horse!"

Judith listened for hours to Uncle Pista's stories of the Hungarian ponies. Most of the time she lived with her grandmother, for her own mother preferred the gaiety of Budapest and Judith's doll-like sister to her elder child. Judith grew up playing with the gamekeeper's sons, who taught her to be fearless and independent. She could hardly wait to finish her lessons and her chores so she could ride Sarga through the forest and across the fields where scarlet poppies and blue cornflowers grew in the native grass that was cut for winter hay to feed the horses that in summer grazed in large pastures. The Hungarians believed in giving the horses as much freedom as possible. Hadn't the Magyar ponies roamed at will across the Steppes of Asia? It was healthier for horses to seek new grass and fresh water than to be penned in a paddock or a box stall.

Unfortunately the pastoral idyl could not last. Hungary, as part of the Austrian Empire, was allied with Germany during World War I. Sarga, along with the other horses on the estate, was commandeered by the Army.

Judith wept in her grandmother's arms. "The Clover is gone! Now we'll have only misfortune!"

"Nonsense!" Grandmere said sharply. "Sarga threw you many times. She was not a good saddle horse and as for your Clover legend, Uncle Pista filled your head with foolishness."

The grandmother did not tell Judith that the old coachman had hung himself in the stable after Sarga had been led away by the soldiers.

In the years that followed, Judith never forgot Sarga nor the coachman's parting admonition, "You must find the Clover horse!"

Judith was an old woman when she told me her story, but her mind was alert and if some incidents were romanticized, what difference did it make? I didn't compare what she told me with her autobiography, of which she made a copy and gave to me. Nor did it bother me when I read articles retelling the adventures that Judith told me. The authors might have quoted Judith or they might have embroidered the story themselves. None of my many Hungarian friends disputed Judith's Clover book. They merely shook their heads over Judith's devotion to her horses; indeed, I had the impression that everything Judith did in America was to save the Hungarian horses as a breed.

Margit Bessenyey, who had known Judith in Hungary, told me that Judith, who was older than Margit, had been admired and envied. This was understandable, for Judith, hardly more than a child, had undergone experiences that her more sheltered friends had not even dreamed of. At her grandmother's request, she had helped the King's Chaplain escape the Communists who staged a revolution when the Triple Entente collapsed. Judith was caught and tortured, but got away by disguising herself as Miss Arizona, to sing and dance in vaudeville.

And I must say that more than forty years later when Judith was telling me her story, she picked up her skirt and, dancing a few steps, sang a refrain of Miss Arizona. Pleased with her performance, she sat down with a naughty smile. "I was very good as a chorus girl."

When the Communists penetrated her disguise, Judith fled again, hiding in peasants' cottages and in hay wagons until she reached the Austrian border, where she caught a train to Vienna. In this city she found her mother, untouched by the Revolution, in a suite in a hotel.

"Judith, what have you done to your hair and your hands? They are badly bruised."

"I caught my hands in the train door," Judith told her mother, whom she knew would not understand what had happened to her. Nor would her mother realize that she was a child no longer. She was a mature woman and she survived as the Magyar maidens had survived centuries before on their journeys across the Steppes with the Clover horses.

★ ★ ★

While she was in her late teens, Judith married a handsome man-about-town and, when the War was over, they moved to Vienna where they led a feverishly gay life, for the Communists made it impossible for them to return to Hungary. One evening Judith visited an amusement park with a crowd of friends where she saw horses for rent. Among these unfortunate animals was an ugly little dun mare with a leather muzzle strapped to her head. The mare refused to move when the owner ordered her to, whereupon he started beating her with a heavy whip.

"Stop!" Judith cried.

"It's my horse!" the man snarled. "I'll beat it if I please!"

"No you won't!" Judith grabbed the whip. "I'll buy it!"

"This devil horse?" the man exclaimed in disbelief. "You'd buy it?"

"Name your price," Judith told him.

The man named a ridiculously large sum and, after some argument, the deal was made.

In her evening dress and fur wrap, Judith led the mare home to her stables where the coachman turned his nose up at the scruffy animal. Judith had to take off the muzzle herself. As the cruel device fell to the floor, the mare's head darted out, her long yellow teeth missing Judith's face by inches. Judith might have been frightened by such savagery; instead, she stood staring, for on the nose of the mare was the Mark of Clover.

This was Sarga, the mare that Judith had ridden as a child on her grandmother's estate. The grandmother had died and the estate had been confiscated, all of which Judith had attributed to the legend. In the morning, Judith had further proof that the legend was more than superstition, for her friends of the night before had driven from the amusement park to a club in the mountains and on the way down had met with an accident in which they had all been killed. If Judith hadn't been leading Sarga home, she would have been killed, too.

Perhaps, Judith thought joyfully, good fortune would come again to her family and to Hungary with the return of the Clover horse and, indeed, conditions did improve, for the Communists were driven out and a Regency estalished in Hungary so that many of the families who had been living in exile in Austria could go back home.

Judith took Sarga to Hungary, although the old mare was so crippled and vicious that she could not be ridden. Not knowing what else to do with her, Judith sent her to a cousin's farm where she was bred several times. At the age of thirty, she died, leaving four colts which, in 1924, Judith decided to send for, since she at last had her own farm. Her first husband had been killed in an accident, leaving Judith with a small boy, and she had married again — a kind mild-mannered man who owned estates in Ujszallas where he encouraged Judith to raise horses. But when Judith requested Sarga's colts, the cousin wrote that the colts had been injured while driving a carriage and had been destroyed.

Dollar, Judith's husband, tried to console her by saying there were other horses.

"But these were Clovers!" Judith cried. "I should have kept them in my stables instead of boarding them at my cousin's. Now the Clovers are gone from Hungary forever!"

Grief-stricken, Judith feared for the future.

For a few years, Judith's pessimism seemed unjustified. She never ceased to regret the Clovers, but she did not brood over a lost cause. She developed a fine stud farm at Ujszallas and rode in international shows in Switzerland and Italy, earning acclaim for her jumping side-saddle. Still, she felt she could do better if she had a really fine hunter.

A cavalry officer with whom she had been riding and who knew Judith was interested in buying a new mount said, "If you want to see a real jumper, Countess, I'll show you one in our quarantine stables. She can leap over a six-foot high fence. We can't keep her in and when she gets out she's the devil to catch — she bites and strikes like a stallion."

Judith laughed. "Show her to me, will you?"

It was feeding time and the soldiers were passing out hay and grain to all but the horse in the quarantine stall. While one soldier opened the door, two others advanced with a pitchfork to protect themselves against the horse that leaped at them, her lips rolled back from her teeth. Forced into a corner, the little mare stood with her ears flattened, her eyes glaring with a reddish light. She had a short neck and a big head, but Judith liked the well-muscled back and hindquarters.

"I like that filly," Judith said. "What is her name?"

"Don't tell me, Countess," the officer laughed, "that you admire our ugly beast! She is a Remount and hasn't been named yet, so we'll call her 'Igezo the Charmer'."

The Captain had his Sergeant ride Igezo, which was no mean feat, for the mare could jump like a bird, but she was herd-bound and had a hard mouth and in the hunting field she killed hounds if she could reach them. At last a regimental board decided the mare should be destroyed or sold at auction. Judith begged to buy her and, for a small sum, became the mare's new owner.

Judith had an instinct about horses and it was because of this instinct that she had risked buying Igezo. From long experience, she knew it was foolish to try to discipline a Hungarian by force. The rougher a trainer became the harder the horse would fight. Iron discipline would not make him behave, but it would make him an outlaw. Some Thoroughbreds would kill themselves from fright or in a frenzy. A Coldblood would lie down and not move and some breeds would be so broken in spirit they would become robots to ride, but a Hungarian, with his strong body, would not give up. He'd kill his trainer if he had a chance.

Judith had an idea that the soldiers had been rough with Igezo so that she mistrusted the sight of a uniform. For a year Judith and her head groom worked with the mare until finally Igezo, who had trusted no man, learned to trust her new owner and, to Judith's delight, proved to be an outstanding jumper.

Years later Judith wrote, "I always said that if I were ever in danger, I could trust my life to Igezo."

So outstanding did the homely little mare prove to be that Judith decided to breed her in the hope that the mare would have a colt that carried her characteristics. Genetics was, as yet, a new science in the equestrian world, but the Hungarians did not need computers to tell them how to breed horses. What Judith did need was Igezo's pedigree in order to register her with the Department of Agriculture. The more valuable Hungarians were listed as purebreds, although this was not correct, for the Hungarians were crossed with the Thoroughbred and Arab, which made them halfbreds, or what is known today as Warmbloods. Igezo, Judith learned from the military, had been raised on her cousin's farm and had been sired by a Thoroughbred steeplechaser. Judith wrote her cousin, who was no more cooperative than he had been. As far as he knew, Igezo's mother had been a work mare. Judith could not accept this, for Igezo did not have the heavy bone or the disposition of a Coldblood, nor the large feet and hairy pasterns. Refusing to give up, Judith went to see the old coachman on her cousin's farm, who gave her information more valuable than anything she had dreamed.

Igezo, the coachman said, was the daughter of Veicze and Veicze had been a daughter of Sarga. Veicze had had a bay filly before she died, which meant that Igezo was the granddaughter of Sarga and that Judith had found the Clover horse she thought she had lost forever.

According to legend, Judith was supposed to be happy foreverafter and for a while it seemed she would be. In 1936, on Igezo, Judith won the tryouts for the Olympics against the official Hungarian team. In addition, Igezo's colts were purchased as studs for the government Remount.

But once again the tramp of marching feet echoed on the cobblestones of ancient European cities and with the armies came the big guns and the fleets of bombers and Hungary was drawn in, as she had been before, to be an ally of the Third Reich. The Hungarian Hussars fought the Russians on horseback in the snows of the Ukraine. Food became scarce. In Budapest buildings lay in ruins. Toward the end of long years of war when the Allies were advancing on Berlin, the Russians were sweeping in a huge tide across Czechslovakia, Poland, Roumania, and Hungary. They came from the East, as had the Mongols and the Turks. Their cavalry were mounted on horses and their supplies were largely hauled by horses hitched to farm-type wagons. In Po-

Judith jumping side saddle at Olympic tryouts in 1936.

land they seized horses as part of their booty; they burned and raped, too, so that those people who could drove their horses into the hills and hid their women and children as they had hundreds of years before.

Judith knew the Hungarian army was defeated and that the Communist sympathizers, who had staged a revolution after World War I and had imprisoned and killed so many of her friends and who had captured and tortured her, would surface again and seize power under Russian protection. Never, Judith decided, would she become a Communist victim a second time. Dollar was somewhere at the Front; Judith couldn't wait for him. She told the people at Ujszallas they could come with her or wait for the Russians. Quite a few chose to accompany Judith, so they loaded their household goods in the clumsy farm carts and carriages, and carts, chickens, pigs, and the horses of Judith's stud started for Austria on the road crowded with refugees.

A worry to Judith was Igezo, who had colic. The veterinarian told Judith the old mare might not make it, but Judith refused to leave Igezo behind. Dripping sweat, the mare tried to keep up with the other horses. As they neared the Austrian border, the Russian guns sounded ever louder and the sky hazed with the smoke of burning villages. Igezo began to stumble, to go down on her knees and try to roll and then to struggle agonizingly to her feet again. When she went down for the third time, Judith knew she could go no further. Kneeling in the dust of the roadside, she cradled the mare's head in her lap.

"We must go on!" the coachman cried.

"Hurry, Countess," the people wept. "The Russians are close behind us!"

But Judith would not move as long as Igezo lived. While she sat there among the weeds and rocks, a young boy approached, cap in hand.

"Gracious Lady, the Russians have set up a roadblock ahead of you. If you continue on this way, you will run into an ambush."

The people crowded around the boy, frightened and apprehensive. "What shall we do?"

"I know a path," the boy said. "I can guide you across the border."

The boy could have been a traitor, but Judith believed him, for it would have been easy for the Russians to outflank them when they had been moving so slowly.

The old mare started to struggle, gave a strangled nicker and collapsed. The poor sweating flanks were still. Judith, fighting against tears, got to her feet.

"Show us the path," she told the boy.

The carts and the carriages, the rest of the horses and the people on foot pressed toward the path, hardly noticing the dead mare in their haste to follow the guide. Only Judith gave one last look at Igezo, knowing that once again the Clover horse had saved her owner from disaster, for if they had continued the way they had intended, they would have been trapped by the Russians. Only Igezo's dying had given them time to be rescued by the boy who, true to his word, guided them safely through the mountains.

Sabrieska, stallion raised by Judith. Ernst Peterson

V
No Grain for Horses in Virginia

Vienna was not the refuge that Judith had hoped it would be. The sour smell of fear and defeat hung over the city that had been divided into zones by the Russians, the Americans and the British. Rations were short, buildings were in ruins, fuel was scarce and refugees crowded into tenements.

Tension was heightening between the Soviets and the Western Allies.

Judith found a farm outside the city in the American Zone where she could keep the horses, but she could not get ration tickets for hay or grain because her horses did not race nor were they draft animals, so she was forced to sell jewelry and household treasures to buy hay on the Black Market. For a while in 1946, the Americans borrowed some of the horses to compete in international shows between the Allies, which eased the feeding problem, but when the American cavalry dispensed with such events, Judith was forced to seek another place for the horses.

Judith was always dramatic and when she was telling us about those chaotic days, she exclaimed, "My God, I did not know what to do! I had sold all I had! The horses were half-starved; I had no money!"

An Allied Colonel helped Judith move the horses to the British Zone where Judith made arrangements for an Austrian farmer to keep them for an outrageous amount. By this time, Dollar had appeared as a refugee and he went to the farm with the horses where he discovered the farmer was a crook who had no pastures or feed. Unfortunately, he'd taken Judith's money and could not be made to return it.

In desperation, Judith went to the Vienna Jockey Club where she was told her only recourse would be to race the horses, which meant taking them into the Russian Zone where the track was located.

For Judith, this was a difficult decision. The horses were starving, their ribs showed, their coats were rough and many had developed bad coughs. They were so hungry they chewed on tree limbs. How could they race? However, there was no alternative, so Judith selected the horses she thought might survive the demands of the track — Sobri and Bori, and a mare named Italia who was a daughter of Igezo. There were others, too. Not only were they required to race, but they had to jump and do dressage for three-day events for the Austrians. The pitifully weak creatures might have given up, but they had the great heart of the Hungarians that would not let them acknowledge defeat. Italia won races and jumping events, as did the rest of them. Some were crippled; some died.

Years later Margit told me, "When you visited Judith in Virginia, you noticed how much grain she gave her horses. She gives them too much and sometimes that makes them too high-spirited to ride comfortably, but Judith cannot forget the days in Vienna when she had no grain to give her horses. She wants them to have all they will eat."

Judith could not return to Hungary, which had become a Communist country under the Russians, and Austria had all the refugees it could manage without offering shelter to a penniless Countess with a herd of starving horses. If she stayed in Austria, Judith was afraid the horses would end up pulling farmers' carts or be sold for meat in the butcher shops. The only hope was the United States.

Judith might not have been willing to trudge the streets of Vienna, to stand in queues for hours to see officials, to beg for help from acquaintances in the Allied Forces, if her own welfare had been the only concern. Her goal was to get the horses to America and that became a possibility when, with the help of an American officer, she was finally granted immigrant papers. She could not, however, take the horses with her; she had to find a place to keep them in their new home.

Again she appealed to an Austrian farmer who agreed to keep the horses for a large sum of money until Judith sent for them. The last of her treasures were sold for feed. How Judith managed, once she reached the United States, to make a down-payment on a small farm she found near Charlottesville in Virginia no one can explain, perhaps not even Judith herself. Nor how she found the money to pay for shipping the horses to New York. Then and later she eked out her meager income by working as a cook for a number of wealthy people and also by serving as a butcher's helper, which meant, most of the time, plucking chickens.

Still, every cent meant one more penny to buy grain for the horses. Out of the sixty-four horses that Judith had when she left Hungary, only six mares, three stallions and four foals lived to make the journey to America and they were scrawny, hungry travesties of their former selves, for the Austrian farmer had mistreated and ill-fed them.

But one of the little foals who laid her ears back at Judith and tried to bite was a filly with the Mark of Clover on her nose.

Such was the foundation of Judith's stud — not much from a commercial standpoint; however, as Judith wrote in her autobiography, "...they were thin, incredibly so, but in their eyes there was the spirit of a breed of horses that...stood by in hardship and died for me and (mine) for centuries."

VI
The Bitterroot Stock Farm

On a wall in the big brick stable, which had a cobbled floor and a cupola on the roof, was a row of pictures. The center photograph was of a grey stallion named Honpolgar IV. The eight mares were of the Shagya, Fenek and Furioso bloodlines.

These were the Hungarian horses that formed the foundation stock of the Bitterroot Stock Farm in Western Montana. They had been brought over from Hungary after World War II by the Army and assigned to various remount stations with the same purpose that selected stock were assigned to government studs in Hungary, Germany and Sweden. Stallions were leased to responsible ranchers and farmers to improve the native breed of horses for their own use, to sell as polo ponies or pleasure horses and to give the Military a better choice of mounts for the cavalry.

Honpolgar, who had been sent to Fort Robinson in Nebraska, had been leased by a rancher in Southcentral Montana whose name was Jim Edwards. Mr. Edwards did not intend to invest a large sum in horses; his idea was to use Hungarians for ranch and cattle work and to sell the few he did not want. The cross between coldblooded mares and Honpolgar was not entirely successful. Hungarians were unknown in the West and Edwards, as he said, could hardly give them away, so when Margit Bessenyey asked if she could lease Honpolgar, Edwards was glad to send the stallion to the Bitterroot Stock Farm.

One of the most historic ranches in the state, the Stock Farm had been founded in the 1890s by Marcus Daly, Margit's maternal grandfather. Mr. Daly had arrived in America as a penniless immigrant, but he had the ability and drive to make a fortune in the fabulously rich mines of the West. He bought a ranch of some twenty thousand acres in the Bitterroot Valley of Western Montana where he raised trotting horses and Thoroughbreds. As an Irishman, the love of horses came naturally to him.

Daly boasted that the timothy hay on his ranch was ideal for blood stock and, indeed, he knew what he was talking about, for hay from the ranch was shipped to Kentucky Thoroughbreds for many years. Clear streams fringed with willows and cottonwoods wound through lush meadows. Horses could gallop across flat grasslands or in the foothills where rocks and timbered gullies taught them to be footsure. In the winter when the mountains across the river glittered like shards of glass, the snow lay on the ground, but the air was dry and invigorating.

Although Marcus Daly never won the Derby, his race horses, among them Tamany and Hamburg, were well-known on Eastern tracks. On the ranch he built a race course, stables, blacksmith shops, houses for his veterinarian, trainers and ranch hands and a brick

Clover mare Mokus with her colt at the Bitterroot Stock Farm. Ernst Peterson

stable with brass door handles for his favorite horse Tamany.

Such was Margit's inheritance on her mother's side, quite different from that of her father, Count Sigray, a member of an old Hungarian family, a horseman and an ardent patriot.

At an early age, Margit showed a talent for riding and a remarkable understanding of horses, which her father encouraged. Her greatest asset was her willingness to learn. She listened to the old coachmen, who were funds of information; and she even attended the Police training school in East Mosley outside London where she cleaned box stalls as part of her training.

When World War II broke out, Countess Sigray stayed in Hungary, which was an ally of Germany, and she remained after Count Sigray was imprisoned by the Germans as an anti-Nazi; she was still on the estate when the Russian Army overran Hungary. Still more remarkable was the courage and skill with which Margit managed the victorious invaders. The Russians respected the young Countess.

Margit could not, however, save her father's horses, which were requisitioned to haul wagons and artillery or for officers' mounts. This meant that horse-raising in Eastern European countries suffered a serious setback. The great estates were plundered of horses, the famous studs in Poland were sacked and Babolna and Kisbir stripped of their valuable animals that had not been sent to a safe refuge in Czechslovakia. Margit saw the race horses, the polo ponies, the brood mares and the stallions—indeed, all the horses that had not gone to the Hungarian Army—confiscated by the Russians. Many of the animals were valuable, for the estate owners had imported English and French Thoroughbred stallions, as well as leasing stallions from the government studs. In other words, numbers of Hungarian horses were identified not only as having come from government studs, but also as having been bred and raised on the estates of prominent noblemen such as Eszterhazy, Batthyanyi and Szechenyi.

When the Americans appeared, Margit served as an interpreter, but when the Allies allowed the Hungarian Communists to take over the government, Margit and her parents had to leave. They went to New York where, after a few years, Margit's mother and father died. Eventually Margit inherited her grandfather's ranch in Montana. Margit went West and when she got there, she rode over every acre of the ranch, including the forested foothills, as well as the pastures and meadows. She noted water rights and hay crops and cattle production and she saw what her grandfather, Marcus Daly, had seen — that the Bitterroot Stock Farm was an ideal place for horses. Only the horses would not be, as Marcus Daly's had been, Thoroughbreds and Trotters — they would be Hungarians.

VII
The Refugee Stallion

The Lippizans were not the only horses rescued from the Russians in 1945 by the Third Army under the command of General George Patton. Other nations hid their national treasures from the enemy; the Hungarians hid their horses, which they valued as highly as crown jewels or paintings.

In 1944, the German and Hungarian governments transferred 600 horses from Mezohegyes—300 from Kisbir and 300 from Babolna—to Germany and Czechslovakia, most of them to the great German Remount Depot at Hostoun. It is possible the Germans had an ulterior motive, since they needed vast numbers of horses for the military and the Hungarian horses were highly rated, but even if they were aware of this, the Hungarian government could not afford to complain.

In 1945, the Germans, haunted by memories of the Russian Front, and the Hungarians by Genghis Kahn, tried to surrender to the Americans, whom they considered the lesser of two evils. In this context, it should be remembered that the soldiers in charge of the horses, as well as the Austrians of the Riding School, were classified as enemies.

According to Judith Gyurky, Alois Podhajsky, Director of the Spanish Riding School, had appropriated the stables at Schloss Arco, a large estate that Judith had hoped to have for her horses. Judith was upset, but the chaos of Europe did not permit old-fashioned courtesies. Podhajsky was as determined to save his horses as Judith was to save hers and, in the long run, not only were the white stallions rescued, but also horses from Kisbir and Babolna by alert American officers who hoped to improve the breed of the U.S. Cavalry mounts.

In his letters, Patton wrote, "We found that the Imperial Riding Academy, which has been teaching high school riding in Vienna since the time of Charles V, had been moved to the vicinity of the XX Headquarters. After lunch they gave us an exhibition of riding and made an address." Patton had his own string of polo ponies at home and was an experienced judge of horseflesh. His orders made it possible for the horses to be saved from the Russians. Good horsemen do not like to see fine animals abused or slaughtered. Patton did not care for the Russians, anyhow, even if they were U.S. Allies.

Not only were the white stallions of the Reitschule rescued, but also a large number of the Arabs, crossbreds and Thoroughbreds at Hostoun. When Patton was told of the Hostoun horses, he ordered one of his Colonels to "get them." Whereupon the Colonel sent a detachment one hundred miles inside Czechslovakia where they rounded up the horses and drove them across the Austrian border practically within sight of the Russians. frustrated and angry, the Russians lodged protests, which Patton ignored, commenting loudly that the horses were worth a lot more and were much nicer than Russian Generals.

Most of the Hostoun horses were taken to an unoccupied German Remount Station in Bavaria where a few months later the American Major in charge of the area distributed numbers of crossbred horses and a few Lippizans to German and Hungarian farmers to work their fields. One can only surmise that the Major was either acting under orders or was no horseman. How many Hungarian horses ended up in stew kettles or had their proud spirits broken dragging a plow is impossible to say.

From the number that remained, Colonel Fred Hamilton, Chief of the Remount Service, selected, in the Fall of 1945, 150 horses to be shipped to the United States. On his second trip in the Spring of 1946, he reported, "I inspected and selected 83 horses for shipment to the United States."

This brief report did not tell the whole story. The six officers who went with Colonel Hamilton to Europe were all volunteers. Two were veterinarians and one was a horse trader who had been outsmarting horse buyers for years. They expected the Germans would try to cheat them and they were right. Their first experience was when a soldier, an ex-ranch hand, told Lieutenant Smith, the horse trader, that he had seen a "peasant" driving some dirty, tired horses who looked far superior to the horses peasants usually had. Smith wasted no time locating the horses. Dirt had been rubbed into their coats and their manes tangled, Smith saw immeditely that they were purebreds worth a lot of money, so the "peasant," an ex-Nazi official, was forced to give up the horses and their papers. At the great Altefeld Breeding Farm they encountered further deception. The war was over and the Russians, hopefully, restrained. The Germans intended to keep their most valuable animals and pawn off their second-best to the Americans, but they underestimated the Americans, who knew all the tricks of horse trading. The Americans were received cordially at Altefeld and the horses were led out of the box stalls, each one provided with papers. The horses were good, but not good enough.

Back of the stables were some old sheds. "I would like to see those sheds," Smith told the Commandant.

Reluctantly, the officer led Smith to the sheds, where Smith found some of the most valuable German Thoroughbreds, including a beautiful mare reputed to be the fastest horse in Poland and also one of the finest young studs in Germany. At another breeding farm, the Americans didn't even look at the horses the Germans were so anxious to show them. They wandered around the poorly kept out-buildings where they located Nordlicht, the most famous horse in Europe and winner of the German Derby.

But all of the horses selected were not Thoroughbreds or Lippizans. Colonel Hamilton was one of few Americans who knew the Hungarians of the Furioso, Fenek, Shagya and Nonius lines and what excellent horses they were for the cavalry, so he selected Hungarian Halfbreds to take back to the States, Honpolgar among them.

Finding freight cars in war-torn Europe was a feat in itself, but finally enough rickety cars were found to ship the horses to Bremerhaven, where the horses, including seventy stallions, were loaded on a Liberty ship. Stalls had been quickly nailed together with boards fastened across the ends to keep the horses confined.

Unknown to the volunteer grooms on the Liberty ship, the Steven Austin, the Press back in the States had discovered the Army was shipping European horses to America and launched a tirade against taxpayers' money going for officers to play polo and ride in horse shows. Unfortunately, politicians listened to newspaper criticisms.

As for the personnel of the Steven Austin, they were too busy to read the papers. A few days out of Bremerhaven, one of the worst storms on record hit the Liberty ship. A huge wave nearly swamped the Austin. The rolling and pitching of the ship flung the horses out of their stalls so that the forward hold was a chaos of screaming, fighting stallions. It took courage for the men to light lanterns and climb down into the hold to separate the powerful animals that kicked, struck and bit, but, finally, after several hours, they had the stallions under control. The veterinarians then had to sew up cuts and treat bruises and take care of the mares who had been injured by falling timbers.

Only the skill and bravery of the volunteers on the Steven Austin saved the ship and its cargo worth millions of dollars in horses from sinking in the storm. But did the men receive any credit? Not at all. They were greeted by another attack from the Press, accusing the Army of bringing home grooms for officers' mounts and leaving high-point ex-combat men in Europe.

The horses were in the country when the Communist Hungarian government asked that their horses be returned and, since the American Army planned to mechanize, some bureaucrats who listened to the Press and saw Russia through rose-tinted glasses felt we should accede to the request. Fortunately, Senator Wayne Morse of Oregon objected and the horses were allowed to remain in America as "spoils of war."

Meanwhile, Jim Edwards, the Montana rancher, wrote the Remount Service to ask for a stallion—a horse he could ride, a good stock horse, "the prettiest you can find." In answer to this request the Remount shipped one of the recently acquired Hungarian stallions to Montana. Jim had never heard of Hungarians and no one else he talked to had, either, but he liked the grey horse the Remount leased him the moment he saw him. He was big with good bone and was well-muscled. His name was Honpolgar IV, which was Americanized to Humphrey, and he was branded with an eagle over a Swastika, the insignia Hitler had ordered put on all German-bred and confiscated horses. According to Honpolgar IV's papers, his sire was Honpolgar—a Thoroughbred, out of Short Holiday—a Thoroughbred, by Duncan Grey—a Thoroughbred. Honpolgar IV's dam was 182 Furioso VII, a Halfbred mare from Kisbir, which meant that Honpolgar IV was a Halfbred.

For a stallion, Humphrey had a very good disposition and in no time at all became, so Edwards said, the best cow horse he'd ever had—sure-footed, sensible, willing and able to go all day on rough terrain. "The only thing he was afraid of," Jim Edwards told me years later, "was water. I imagine that was because of the time he was nearly drowned on the Liberty ship coming to America."

"You know," Jim smiled as he reminisced, "Humphrey never knew he was a horse; he thought he was a person. We didn't keep him in the pasture but fixed up a stall for him in the barn by the house and let him graze in the yard. He used to get in the tool shed and carry off my tools and hide them."

Helen Edwards joined in. "He followed Jim and me like a dog and one day when I came home with some groceries, I set them on the sink and went out to feed the chickens. I couldn't see Humphrey anywhere. I was really scared. Had he opened the gate? He was learning to open doors and gates. I searched everywhere and then, out of the corner of my eye, I saw Humphrey's horse face in the kitchen window. He had opened the front door, gone through the living room and bedroom into the kitchen and was standing at the kitchen sink

Honpolgar (Humphrey) as a stock horse.

eating potatoes from the grocery sack. We hadn't had him awfully long and I didn't know how I could get him out of the house. I let him finish the potatoes and then I said, 'Come, Humphrey,' and he turned and followed me through the bedroom and the living room out of the house."

In 1949, Jim Edwards received a letter from the Remount saying the service was being discontinued and the horses were to be sold at auction. An officer would come to Montana to collect Honpolgar. Jim called one of the officers at Fort Robinson whom he knew. "You can send someone to get Honpolgar, but when he gets here, he won't find me or my horse."

The result was that the Remount officers figured it would cost more to send someone to Montana than it would to let Edwards have Honpolgar, so they sold the stallion to Jim for a small amount.

When the Remount sold out, Margit Bessenyey heard there were some Hungarian mares to be disposed of. Wasting no time, she asked a refugee Hungarian veterinarian to buy the mares and ship them to Montana. Of the mares Margit purchased, Gazal II had been bred at Babolna, Furioso II at Kisbir, Shagya XXV at Babolna, Szeplak at Kisbir, Fenek VI at Kisbir, Formos, listed only from Hungary but bred as part Thoroughbred, Shagya XXVII at Babolna and Furioso VII at Kisbir. These were bloodlines Margit wanted, for horses of this kind had produced cavalry mounts, polo ponies, racing stock, hunters and pleasure horses. Both Judith and Margit decided they would emphasize the Hungarian as an all-around horse. The question was where to find a stallion that would suit Margit's breeding program.

The model Hungarian, to Margit as it was to Judith, was a medium-sized horse of fouteen three to fifteen three, short-backed, muscular with a neck well set on a sloping shoulder, a deep chest and refined head, the eyes wide apart, and colors predominantly brown, bay and dappled grey, although some of Judith's Clover horses had been duns, as were many Mongolian ponies. Also, the Clovers were apt to have shorter legs and larger heads than the Hungarians, which had more Thoroughbred and Arab breeding.

The two women had good reason to believe in this type of Hungarian. In 1911, Dr. Van Slatin, the Master of Imperial Horse in Austria, said, "We believe the horses bred in Hungary are the best Cavalry mounts in the world. They are not too large and have a great proportion of Arab blood in them."

After World War II, an Italian specialist in equine subjects, who favored the Thoroughbred in the Hungarians, wrote, "Like most other European nations, Hungary has seen its horse population declining during the present century, but its breeders have effectively reversed the trend over the last twenty years, producing some of the most beautiful horses in the world."

Height was not, so Judith and Margit felt, of paramount importance. It never had been in the history of the Hungarian since the original pony had become a moderate-sized animal with cross-breeding. Proponents of the Oriental bloodlines, like their kind who had been so vociferous in 1815, pointed to the English, whose big Cavalry horses had proved unusable in the relief of Khartoum, while the little Arabs of their Egyptian allies had endured the campaign without difficulty and again, in Africa, the English, who seemed slow to learn, lost their big horses to disease and lameness, while the Boers ran rings around them on their ponies. Not until the English shipped American mustangs to Africa could they prove successful against the Boers.

In the *Remount Magazine* of November and December 1946, an editorial read, "Perhaps the most valuable horses that were imported, however, if we take a long-range point of view, were the Hungarian Halfbreds.... In a matter of size, they breed far more truly than do most breeds. Almost every Halfbred stands between fifteen and fifteen three and most of them will weigh very close to eleven hundred pounds. In confirmation, they are big horses on short legs, deep through the heart, broad of shoulder, with great bone...short cannon bone and good feet. They are fine, yet powerful."

Of the mares Margit bought from the Remount, all except the Arabs from Babolna had a great deal of Furioso breeding. The first Furioso, as we know, had been a Thoroughbred imported from England in 1840. From 1841 to 1851 he produced ninety-five stallions. At intervals the Furioso bloodline was strengthened by additional English stallions, including the well-known North Star, who, according to Miklos Jankovich, was a trotting "roadster" bred in East Yorkshire. Margit's mares were of Thoroughbred or Arabian breeding, which meant she could not use a stallion of the same bloodlines without risking inbreeding. Fortunately, Margit heard about Honpolgar and got in touch with Jim Edwards, who agreed to lease the stallion to Margit in return for a certain number of colts.

This was a busy time for Margit. She married an Hungarian, Baron Bessenyey. They did not live permanently in Montana, but spent part of the time on an estate on the Potomac that Margit bought with the idea of having an Eastern base for the horses. Actually, the estate was an excellent investment, a pre-Revolutionary War plantation with a classic white house and rolling green pastures. Another advantage of the Maryland farm was its proximity to Judith Gyurky's farm in Virginia.

Montana was the location of the stud farm where the

Honopolgar at Bitterroot Stock Farm. Ernst Peterson

horses were bred and raised and where they received their preliminary training as colts. Margit, who did not have the romantic attachment for her grandfather's ranch that she had for Hungary, boarded up the big pillared house to save on taxes, locked the wrought-iron gate of the long, maple-lined lane to the house and made the lane and gardens around the house into pastures for the horses. As time went on, she built more rail fences and a stallion area with shelters and several acres for each horse to exercise. Barns were also repaired and the weeds plowed out of the mile race course.

Ranch help was a problem, as it has been for many years, but Margit managed to find a reliable foreman and a man to take care of the horses. The latter was Willy Couture, a professional jockey who had been accustomed to hot-blooded horses. The cowboys who worked on the cattle part of the ranch viewed the Hungarians with distrust, judging them too skittish. Jim Edwards might have been able to make a stock horse out of Humphrey, but Jim was an able horseman and Humphrey an obliging animal.

The cowboys never rode the Hungarians. Indeed, the cattle oufit and the stud farm were separate entities, each with its own foreman, and Margit overseeing the entire ranch. No mistake about it, Margit was the boss. The men respected her because she not only knew her business, but she was willing to learn and to listen, she was a good horsewoman and she was interested in the men's families. To show that she was accepted, the men called her by her first name.

Humphrey arrived from the Edwards ranch on the 11th of June 1954. Jim's father-in-law came with him. The picture on the stable wall shows Humphrey as an ideal Hungarian, with short back, broad chest, good bone and sloping shoulder, the muscles well-developed over the back and loin, the head well set and the line

Honpolgar's first colt at Stock Farm. Ernst Peterson

between jawbone and throat clean-cut. In color, he was dappled grey. His forehead was broad and his eyes large and intelligent. He weighed twelve hundred pounds and stood fifteen two.

Willy Couture, who was in charge of the Hungarians, kept a notebook of the dates the mares were bred, when they foaled, when the colts were weaned and their characteristics. Willy was a true horseman, which meant he had the skills as well as the instincts.

One of the mares who foaled was Furioso II, dubbed Nibbles, who had a filly that Willy called Nell. By the end of June, foals raced and played in the pastures, watched by indulgent mothers.

In his notebook, Willy wrote, "New Babies Reactions." This was when the colts were being halter-broken.

"Peter by far the most wild of all.

"Alsab nice disposition and doesn't fear people.

"Mickey bit me. Later found he had broken a bone in my arm.

"This group of babies as a whole by far the most wild and difficult to start handling. If not very careful, sure to be kicked. Very unusual for babies to kick at a person.

"After weeks and days with the babies, this morning was the first time they greeted me."

Half a century after Marcus Daly had bought land in the Bitterroot, his ranch had become a Hungarian stud farm.

At the moment, it did not concern Margit that Western Montana was remote from the big horse shows and that trainers experienced in dressage, three-day events and jumping were hard to find in the state.

VIII
The Cooksley Stud Farm in Nebraska

For over a thousand years, fate or luck—call it what you will—played an important part in the survival of the Hungarian horses. At the crucial moment, dedicated individuals appeared to save the animals from extinction. Judith attributed it to the mystique of the Clovers. Or could it have been Acts of God?

Whatever it was, it was once more manifested when Jim Edwards brought the Steven Cooksleys to the Bitterroot Stud Farm to see Honpolgar and the mares Margit had bought from the Remount, a visit that was to be a turning point for the Hungarian horses in America. Without the Cooksleys' help, Margit and Judith would have had a difficult, if not impossible, task raising their horses.

From that time on, Margit, Judith, the Cooksleys and, to a lesser extent, Jim Edwards formed an unofficial association that put the breeding program on a businesslike basis. Margit leased Honpolgar in return for Edwards's pick of the yearly colt crop. Cooksley and Margit, as well as Judith, exchanged stallions and mares and, in some cases, Margit bought horses from her associates.

Steve Cooksley was a real horseman who had grown up in Nebraska and, when he was married, raised a Quarter Horse-Thoroughbred cross. His business was successful, but Cooksley was not satisfied with the Quarter Horse, the most popular riding and show animal in the West. In 1946, he went to a Remount dispersal sale to buy some Thoroughbred mares. As it happend, three of the mares had been bred to Hungarian stallions—two to Fenek V and one to Furioso II. Like Edwards, Cooksley knew little about the Hungarians, but when the colts from the three mares were old enough to be ridden, the Cooksleys were amazed at their stamina and intelligence.

As Wanda Cooksley, Steve's wife, said, "we put a group of horses in the corral and ran them around and around and it took very little time to decide upon which back you would like to be riding."

So the Cooksleys sold their Quarter Horses and began raising Hungarians, a risky move, for Quarter Horses were an established breed, while Hungarians were unknown.

The Hungarian project was fraught with difficulties. By 1948, it was as difficult to locate the horses that had been sold at the Remount dispersal sale as it was to find a penny in an oat bin. Fortunately, Cooksley's brother was a Colonel in the Remount and he was able to get sales catalogs and addresses of people who had bought Hungarians. In this way, the Cooksleys located in Nebraska a stallion of the Fenek line, to whom they bred many of their mares. They finally persuaded the stallion's owner to sell him, but before the Cooksleys could pick him up, he got into the turkey feed and died. A second stallion, in Wyoming and also a Fenek, that the Cooksleys wanted to purchase was poisoned by loco weed.

This left Honpolgar, as far as the Cooksleys knew, the only stallion in the area, but since Margit had already contracted with Edwards to lease him, he could not stand at stud in Nebraska. The best the Cooksleys could do was to arrange with Jim Edwards to have one of Honpolgar's colts out of a 158 Fokas mare for a future stud. The next year or so the Cooksleys bred their mares to an Arabian stallion and an Argentine Thoroughbred. They also bought two fillies from a man in Ft. Reno, daughters of a 6 Conul mare and a Formas stallion, and a third mare in Oklahoma, the daughter of an imported Hungarian mare.

When he was two years old, the colt that Cooksley had liked and that Jim Edwards had claimed as his share of the Bitterroot Stud Farm colt crop, arrived in Nebraska. H. Kallo 131, nicknamed Bill, was by Honpolgar out of an imported mare, 158 Fokas. Bill was a bay and Cooksley considered him "one of the most prepotent and best stallions" he had ever seen. In time, Bill went to the Bitterroot and Honpolgar went to Nebraska, where he immediately became the star of the ranch.

"Honpolgar," Wanda Cooksley wrote, "was a lovely horse.... It was a delight to watch Steve ride him, for he had such a long and free vigorous stride. His disposition was as near human as possible. He bugled everyone into and out of the yard from his pen just south of the old barn. He was better than a watchdog."

Big John, Honpolgar's son out of Hungarian Lulu, one of the Formas mares bred at Ft. Reno, had his father's good disposition. A hunter type, he stood about sixteen hands. When the Cooksley's herd showed a preponderance of Big John mares, Steve Cooksley sent the stallion to the Bitterroot and received in return a grey horse by Honpolgar out of a mare that Margit bought from Temple Smith in Chicago. The mare was one of a number that Mr. Smith had imported from Hungary before he decided to concentrate on Lippizans. The grey stallion, named Hadur, was never broken to ride, but on his own he had a beautiful, long, extended trot and was a handsome animal.

In the 1980s, the leading stallion at the Cooksleys' was Hungarian Taltos by Magyar Brado, raised by Judith out of Hungarian Nora, whose sire was Bill.

IX
The Toughest Test of Man and Horse

Margit's concern was not only for the horses, but also for the riders. A person might own a spirited Hungarian or a cayuse, but if he did not take proper care of his animal and if he did not know how to ride, the horse would not perform to the best of its ability.

Most of the problems in the West, I felt and Margit agreed with me, resulted from a borderline economy. Ranchers couldn't afford veterinarians; most did their own shoeing and few wormed their horses. Also, they could not spend the time needed to properly train their horses.

"The best way to learn about horses," Margit said, "is to train for a Long Distance Ride. We can put on a fifty-mile Ride on the Stock Farm and hold clinics in the spring to teach people how to feed their horses and how to ride. At the same time, we can train for the Tevis."

The Tevis is the hundred-mile Endurance Ride across the Sierras in California, often called the toughest test of horse and man that exists. Margit wanted to ride a Hungarian to prove the endurance of the breed. If she rode herself, she would show her faith in her horses. A woman trainer in California, who was preparing a number of Hungarians for jumping and cross-country events, would accompany Margit and so would our daughter, Tex, on her Hungarian gelding, named Flower Child.

One of the local veterinarians volunteered to lecture at our weekly get-togethers, held on the lawn beside the brick stable. To our surprise, nearly fifty people showed up at the first meeting. Taszilo, the horse Margit was riding in the Tevis, was one of the models used by the vet.

"If you want a show horse," the vet said, "you need a pretty animal that likes to go around a ring, but if you want a horse for long-distance riding and cross-country events, you want a strong, aggressive animal that isn't going to quit on you."

Such an animal had been the Hungarian that in 1892 had won the famous Berlin to Vienna Ride between the Austro-Hungarian military and the Germans. The horse was a mare and she covered the three hundred fifty miles in seventy-one hours and twenty-seven minutes. Since then, rules for long-distance Rides had been made stricter; tough periods of training were mandatory and horses were checked before and during the Rides by veterinarians who disqualified animals that were having difficulties.

Margit had entered her horses in Rides in Vermont, Florida and Virginia, where they had placed well. People had been impressed, but the supreme test would be the Tevis.

To novice riders who would compete in future Bitterroot Stock Farm Rides, the veterinarian explained basics that applied to Hungarians, as well as other breeds. Get a horse with sound legs and feet and a straight way of going. Have him shod every six weeks by a good blacksmith. Make sure he is immunized and wormed. Feed him fresh hay, timothy or timothy and alfalfa and see that he gets the proper vitamins and supplements. Exercise him regularly, using an English or stock saddle with adequate blankets and remember that the best gait to cover ground is the extended trot. Above all, learn to take the horse's pulse and respiration, one of the best ways to judge an animal's condition, although other factors are important, such as dehydration, lameness and fatigue.

Checking respiration, the veterinarian explained, can be done by looking at the horse's sides or holding a hand to the horse's nostrils. At rest, respiration averages between seven and twenty-five breaths per minute. If it's fifty-five or over and the horse continues to heave and gasp, there might be a problem. Usually, the horse heaves a few times, then takes a deep breath and his breathing returns to normal.

The pulse can be detected in four places—by the artery on the inside of the jawbone, in the groove of the tail near the root, by the digital pulse below the fetlock and by the pulse in the chest. A normal pulse of a horse at rest is between twenty-eight and forty, depending on the age and breed of the animal. To take the pulse, count the heartbeats for fifteen seconds and multiply them by four, which gives you the pulse per minute. Naturally, a horse who has been trotting will not have a normal pulse. If, however, its pulse and respiration decline after ten or fifteen minutes to a satisfactory rate, the horse is in good condition.

"The hundred miles will tell the tale," Margit said.

Faults did show up in some of the novice horses; however, the first time we had the Bitterroot Trail Ride our one entry for the heavyweight division died of worms a week before the competition. The next year the horses were better trained and a higher grade than the animals the first year, until, after a number of years, the horses and riders could have competed in any Ride in the United States. Except, perhaps, for the Tevis, which demands a horse and long hours of training most people cannot afford.

The Tevis is a hundred miles in twenty-four hours, which meant that the year the Hungarians were entered the Ride began at dawn, continued during the day and night and finished at sunrise. The start was at Squaw Valley where the track climbed twenty-seven hundred feet to the summit of a pass. The trail was narrow, rocky and dusty and when the sun came up, it was hot. At Cougar Rock the trail led straight up a cliff, with drops of hundreds of feet on both sides. Flower Child and Taszilo looked at the obstacle and, without faltering, scrabbled up the precipitous slope. When possible, the horses had to trot, with the mountain rising on one side and dropping hundreds of feet to the canyon on the other side. A number of horses began to falter; some riders clutched their saddle horns in exhaustion. A few horses were eliminated by the veterinarians at check stops. The Hungarians, however, passed all the fitness tests. Shortly beyond the first vet check, the trail, narrow and eroded, switchbacked more than two thousand feet into the middle fork of the American River. To look over the side of the trail, barely wide enough for a horse, was to risk giddiness. Across a suspension bridge and up two thousand feet from the canyon the temperature hovered over ninety degrees. One rider collapsed by the side of a trail. A girl sobbed against her horse's neck, "I can't go on!"

Some horses developed colic; others had thumps.

After an hour's rest among the trees of a small settlement, the riders went on in the gathering darkness.

Margit riding the Tevis.

There was no moon, so the horses had to feel their way on the trail. Tex became separated from the others and, in the blackness, rode across a bridge sixty feet above the water, that had no guard rail. Flower Child snorted and picked his way carefully, but he did not object to going where Tex asked him to. On the opposite bank, the trail followed an old railroad grade. Suddenly Flower Child stopped. Tex tried to knee him forward, but he would not move. Switching on her flashlight, Tex saw a washout that, if Flower Child had not sensed it, would have resulted in a nasty fall.

None of the Hungarians won the Ride, but they were not trying for the trophy, only to prove they were foot-sure and trail-wise and knew how to take care of their riders.

After the Ride, when I observed Tex and Flower Child, I could sense a bond between them similar to that of two people who have been through a life-threatening situation and, by depending on each other, managed to survive. That bond, which horsemen know, has been forged through the centuries from the time the daughter of the High Priest was saved by her Clover stallion from torture to the time Alexander galloped Bucephalus across the deserts of Asia Minor to the years when Genghis Khan and the Seljuk Turks raced over the Hungarian plain. The rider who depended on a weak horse did not live; only those survived who rode a horse of great heart, loyal unto death.

X
Establishing a Registry

At the Tevis, people said, "A Hungarian? What's that? I've never heard of a Hungarian."

Margit and Judith were doing their best to inform people about the horses. That meant, first of all, to establish the breed, which meant having a registry. In the case of the Hungarians, this was not easy because records had been lost or destroyed and the government was occupied with more pressing matters than horses following the 1956 revolt. Margit could not expect cooperation from the People's Republic of Hungary when she had gone back to Europe and helped people escape from Hungary during those troubled times.

Judith, of course, had records of the horses she had brought to America so that when she sent a stallion or mares to the Bitterroot, their records went with them. The Remount horses had papers, too, due to the Germans, who are a methodical people, but many stud books were lost when the big Hungarian estates were confiscated.

In order to establish a registry, Margit hired Dr. M.E. Ensminger, President of Agri-Services Foundation, one-time chairman of the Department of Animal Science at Washington State and a lecturer to scientific groups in Moscow and Peking. At the same time, Hannah, the Hungarian secretary, spent months tracing the bloodlines of the Cooksleys', Judith's and Margit's horses.

In 1966 the Hungarian Horse Breeders Association held a meeting at the Stock Farm, presided over by Dr. Ensminger. Immediately there were problems. Dr. Ensminger explained that the National Society of Livestock Record Associations required that all foundation stock pass an inspection for registration over a period of seven years before being recognized. Also, there was a limit to the amount of Thoroughbred or Arabian breeding permitted in the Halfbreds.

Cooksley and Edwards were disgusted; Margit and Judith were furious.

"Only a knowledgable Hungarian would be qualified to inspect Hungarian horses!" Margit declared.

Edwards and Cooksley felt such inspections would cost too much.

"The Hungarian is an old breed!" Judith cried. "Their registry goes back hundreds of years! They do not have to be judged like Quarter Horses!"

All four breeders rejected the limitations on Thoroughbred and Arab blood that could be introduced into the Hungarian. In Hungary there were no such restrictions. The infusion of Arab and Thoroughbred blood was left to the discretion of the breeder.

Then, Ensminger argued, the Livestock Association people would point out that no breed was permitted the infusion of two bloodlines without restrictions and the Hungarians would be classed as a three-breed cross, rather than a Purebred.

BREEDERS MANUAL

HUNGARIAN HORSE ASSOCIATION

"Purebred Hungarian horse breeders control the destiny of the breed; hence, it is important that they chart its course well and that they maintain high standards.

"The material for this manual was prepared by Dr. M.E. Ensminger under the sponsorship of Mrs. Margit Sigray Bessenyey, Hungarian Horse Breeder."

The Livestock people, Ensminer tried to explain, were not trying to be difficult; they were only trying to protect the legitimate breed associations from "groups which fall by the wayside within a short period of time" and "maverick groups who all of a sudden think they have a new breed."

At that time it was impossible to foresee that within ten years the European Warmbloods would be introduced into the United States and that Hanovarians, Swedish Warmbloods, Dutch Warmbloods and Trackeners would command big prices as jumpers and dressage animals. The Hungarians fit the definition of Warmbloods, with the added distinction of being recognized by many experienced horsemen as the best all-around light horse in Europe, with generations of careful breeding behind them. In Hungary they were recognized as Purebreds; that is, a definite type. It was a pity the Hungarians could not have profited from the future popularity of the Warmbloods. If they had been able to do so, their story might have been far different.

Margit's temper exploded more than once. Judith was more restrained, but both women bitterly resented the inference that their horses, with generations of breeding, should be considered ordinary crossbreds. Nor did it help that a member of the Board selected by the Stockgrowers Association to pass on the Hungarians was a representative of the Pony of the Americas. Another was from the Shetland Pony Club and a third from the American Quarter Horse Association, which had not been established until 1940.

However, the authority of the Stockgrowers Association could not be disputed. If they did not pass on the Hungarians favorably, raising them would be considered a hobby and the owners would be taxed accordingly. So, as Dr. Ensminger and Margit's legal adviser declared, it would be wise to compromise. Margit, supported by Ensminger, appeared at a meeting in Chicago and some months later the seven-year restriction was modified. The imported horses were listed as foundation stock, a request for registration was printed, as well as registration forms of the Hungarian Horse Association, and the Association incorporated with Cooksley as President and, later, Judith, and with Margit as Secretary-Treasurer.

Thus, one obstacle, and an important one, was overcome.

But there was another problem that was common to most horse breeders and that was particularly acute to the Hungarians because there were so few in the United States, and that was inbreeding. The dictionary defines inbreeding as, "To produce by the continued breeding of closely related individuals." Other sources say that inbreeding is the crossing of animals who are more closely related than usual. Line breeding and close breeding are forms of inbreeding, but to a greater degree; that is, in close breeding a sire could be mated to a daughter or a brother to a sister. In line breeding, the animals are not the near relatives they are in close breeding and the purpose is usually to produce a colt with the speed and stamina of a champion sire. Thoroughbred breeders practice line breeding and the Hungarians have been line-bred for generations, which is the reason they remain true to type despite the infusion of Thoroughbred and Arab blood.

XI
Linebreeding— For Experts Only

Close breeding is seldom practiced among the various accepted breeds because of the danger of deformed colts and undesirable mental characteristics.

Dr. Ensminger, however, says that from a biological point of view, line breeding and close breeding are the same thing. The advantages are that in the former an animal is apt to be more prepotent; that is, a stallion could pass on his characteristics and, in general, the bloodline of line-bred animals would be desirably uniform in appearance and character.

Wanda Cooksley wrote, "We have a friend who is a livestock genetics specialist... who has been interested in the horses and has advised us on the line breeding aspects of the breeding program. One of the problems of line breeding is that undesirable defects will be become apparent and more numerous. The amazing thing about these horses (the Hungarians) is that in spite of close line breeding, they have not produced defective horses and because they are so line-bred, they breed very true to type within their breed and are prepotent (their offspring are most likely to be more like them than the outcross) when bred to an outside line."

However, Dr. Ensminger and other knowledgable equine specialists caution that only experienced stud managers should practice line breeding and they should be financially situated so they could afford to dispose of a deformed colt, should one appear in the herd.

"Honpolgar," Wanda Cooksley wrote, "is very close in all of the extended pedigrees, but," she continued, "Honpolgar was a horse with enough character and style to merit his position as virtually the foundation stallion of the horses bred at the Bitterroot Stock Farm and the Cooksley Ranch."

Honpolgar, or Humphrey, enjoyed the prominent position he occupied. In the spring Jim Edwards trucked him over the mountains to the Stock Farm and in the fall he brought him back to the ranch, where he became a cow pony again.

The first time Jim took him to Margit's, he found Humphrey asleep in the barnyard with one of the cats dozing on his back. He loaded Humphrey in the stock truck and started out. When the road was long and empty, Humphrey lay down in the truck, but when they approached a town, he stood up so he could see himself reflected in the store windows, whereupon he would neigh loudly. Then, at the edge of town he would lie down again until the next town, where he repeated the performance.

After several years Margit added another stallion to the Stock Farm program — a purebred Arab named Pilot who came from Idaho. There is some uncertainty as to Pilot's origins, but in the list of horses imported by the Remount there is a purebred Arab, Pilot, standing

Pilot. Purebred Arab stallion. Ernst Peterson

fourteen two, a grey colt foaled in 1943, listed as Number Eight, and by Trypolis out of Zalotna. The description fits Pilot at the Stock Farm.

In the Breeder's Manual for the Hungarian Horse Association, Judith Gyurky wrote:

"As a result of over 120 years of line breeding, the Furiosos pass on their qualities and appearance to an exceptional degree. Baroness Bessenyey has a concentration of Furioso blood in several of her families.

"Siglavy-Gidran was the originator of the now existing Gidran family. Gidran was a dark chestnut with perfect confirmation, with the exception of rather low withers. This fault, after one hundred fifty years, is still recurring in the family. Through 150 years the Gidrans were crossed with Thoroughbreds to get the size and elegance necessary in modern horses. They produced very hardy horses with a lot of jumping power and lively but resentful dispositions. The regional breeders (farmers) did not like them, primarily because of their poor dispositions and chestnut color. The Gidrans would not tolerate poor horsemanship or brutal treatment...(some) countries characterized the Gidrans as vicious and unreliable. This thinking was reflected among Hungarian breeders, too. Hungarians have a saying, 'Red woman, red horse, and red dog are never good.'

"Countess Judith Gyurky has the best Gidran blood in some of her horses."

In the Manual, which was written by members of the Association, we are told that the Arabs were used to regenerate the herds of horses that became too inbred or weak or heterogenous.

"Arab stallions performed wonders in such breeding programs. In one or two generations they unified the type of horses and stamped out softness and other faults.

"The Arab bloodlines in Baroness Bessenyey's Stud Farm," we read, "were the best available... Countess Gyurky has also a strong strain of Bablona Shagya in her herd. The Shagya, having more size than the average Arab... measured up to fifteen two and fifteen three."

"Kisbir," the Manual says, "was the cradle of the fame of the modern Hungarian horse." As Spencer Borden reported in 1911, Kisbir specialized in Thoroughbreds and English Halfbreds, so-called because of the infusion of English Thoroughbred blood.

"The reason for the excellence of the Kisbir horses was that the finest of Thoroughbreds were used for breeding — horses like St. Simon bred in 1881 by Prince Batthyanyi, who won the gold cup at Ascot; Kisbir, who in the 1890s won the English Derby and the French Grand Prix; and Kincsem, the amazing mare foaled at Kisbir, who won races in Germany, the Grand Prix at Deauville and the Goodwood cup in England.

"The confirmation," the Manual continues, "size, and straight gaits were strictly observed and caused by a careful selection in breeding. Baroness Bessenyey has the best representatives of Kisbir blood in the United States. Countess Gyurky has, like all Hungarian breeders of fine horses, Kisbir strains in her breed of horses, also."

So the Hungarian horses, a recognized breed, were established in Judith's Port-A-Ferry Farm in Virginia, at the Cooksley ranch in Nebraska, and at Margit's Bitterroot Stock Farm in Montana.

The Origin and Development of the "Nonius"

by

Major Janos Devenvi

Commander of the Royal Hungarian
Herd Breeding Unit in Mezohegyes, Hungary

XII
Nonius, the Dependable Animal

Margit and Judith had few Nonius bloodlines in their horses. They preferred Furioso, Gidran, Shagya and Thoroughbred; however, the Nonius was popular in Hungary, especially among the peasants, who used Nonius horses to pull wagons and for work in the fields. Margit and Judith felt the Nonius was too heavy and lacked refinement and agility; yet the Nonius was, in many respects, an admirable animal. A resume of its history tells us much.

In 1810 in France from the herd of Rossieres in Normandy, a male colt named Nonius was born to a mare of the third generation from an English stallion and Orion, an English Thoroughbred.

The sudden appearance of an Austro-Hungarian Cuissarier unit (the Hussars) resulted in the "appropriation" of Nonius and several other horses, who were transferred from France to Mezohegyes in Hungary.

"Apparently," the Major wrote, "the Captain of Hussars must have had good eyes and good sense concerning domestic horse breeding because along with Nonius, numerous other excellent colts were taken away.

"The culture and climate (of Hungary) were perfectly congenial to the Nonius and he created a line of such superior species that it cannot be duplicated in any part of the world."

Major Devenvi, it is to be feared, was even more partial to the Nonius than Colonel Hedad at Babolna was to his Arabs. It is interesting to note, however, that the Major always wrote of the Nonius as Coldbloods.

"At the present time, when the requirements of horse breeding have undergone an essential metamorphosis, the value of the Nonius progeny has been greatly increased because, as types of endurance, they surpass all other coldblooded stock, even on the heaviest ground.

"The Nonius breed runs easily in cart or coach; it is light and well-timed under the saddle and it is excellent for fox hunting."

The following examples may be cited. At Mezohegyes, firewood had to be hauled in from the forest. Major Devenvi, remembering how the small Arabs hauled heavy loads as far as twenty kilometers, felt he might employ a pair of big Belgian stallions he had at the Stud Farm to haul wood, but "the giants" refused to pull, especially uphill; whereupon, in desperation, the Major turned to two big Nonius horses. They were hitched to a cart and pulled the loads of firewood with no trouble at all.

It was an established fact, the Major said, that the Nonius could haul over long distances, through mud and up and down mountains, even though they were not raised in rough country.

The Nonius, the Major insisted, also excelled as hunters. Two mares, Nonius 510 and Nonius 493, were ordered to Budapest to the Fox Hunting Association to be tried as hunters. When the Master of the Hunt, Prince Egon-Thurn-Taxis, saw the mares, he commented in disgust that they looked like truck horses, but when his own mount went lame on the rough, muddy ground, he had to change to Durcas Nonius 510. The Prince led the ensuing runs on the mare and when they were finished, he galloped the mare twenty-two kilometers. Dismounting, he stood back to admire the mare, saying, "Marvelous! Marvelous!"

As a result, the two Nonius mares became hunters for the Association and were often ridden by the Prince.

From the description of the mares and from the stud books, Nonius 510 and Nonius 493 must have resembled Irish hunters, a cross between Thoroughbreds and draft animals.

"If the Nonius had not 'accidentally' come to Mezohegyes," the Major wrote, "he would never have become the founder of a noted and outstanding line. His name would have been lost in the obscurity of some Norman village, where he would have conveyed over stony roads the wares of some peasant."

In the ensuing years the Nonius was crossed with Arabs of the Siglavy line, with Hungarian Halfbreds, with Lippizans and with Thoroughbreds. In the Nineteenth and early Twentieth Centuries, three Nonius stallions became well-known—Nonius XXIX, Nonius XXII and Nonius XXXVI. The stallions exemplified the best of the breed. Interestingly enough, Nonius XXXIX was third generation from a Nonius-Siglavy crossing.

The Nonius enjoyed a favorable breeding program up to World War I and the first Communist revolution, but since that time, the Major wrote, the breeding of the Nonius has been waning, "due partly to lateral and previously promoted Thoroughbred breeding and, furthermore, to the deterioration consequent upon the

Nonius stallion. Budapest.

decreased food supply during the Roumanian invasion. The characteristics of the Nonius began to alter profoundly. The belt measurement lessened, the extremities stretched and the heart area shrunk. In fact, the Nonius was divested of the intrinsic characteristics and shape of the ideal type exemplifying beauty and power.

"The opinion today is that the Nonius must preserve its integrity of species, with its attributes of power, endurance, capacity, beauty and steadiness, all of which must be accented.

"It happened at times that the large ram head was erroneously interpreted; whereas, it is not a fault, but a characteristic property."

This ram's head, which was a large, coarse Roman-nosed appendage, was seen not only on some of the Nonius horses, but also on some Lippizans, such as the ones that drew the carriage of the Emperor of Austria-Hungary. Most people thought it was homely, but Major Devenvi, forever loyal to the Nonius, insisted that "the strong head is associated with nobility, meaning, besides power, placidity. And what other quality is more important in a draft and general purpose horse that is eminently desirable?"

Hungarian horse at California horse show. Ed Lawrence

XIII
The Versatile Horse: Driving and Grand Prix

To publicize the Hungarians, Dr. Ensminger recommended a brochure and entering the horses in shows, particularly in jumping and three-day events. The Cooksleys had already entered some of their horses in Midwest shows, but the popular breed continued to be the Quarter Horse and judges were inclined to favor that type, which discouraged the Cooksleys and did little to further the cause of the Hungarians. Judith did better because she was in horse country where she belonged to a Hunt and could enter her horses in the local shows.

Jim Edwards, who was a working rancher and had only a few Hungarians, had neither the time nor the money to attend shows and to advertise. To Jim, Humphrey was his cow pony, although with a little training, Humphrey might have been shown. He loved crowds and his proudest day was a Fourth of July when he led the parade in Hamilton, the town adjacent to the Stock

Farm. Humphrey's rider bore the flag and the grey stallion bowed his neck and pranced as though he carried the President himself and did not have a Swastika branded on his shoulder.

Humphrey was gaining a certain amount of fame as a refugee from the Nazis. Articles about him appeared in local papers and he even had a fan letter addressed merely to "Honpolgar, Care of Edwards, Manhattan, Montana."

But Humphrey's fame was not, unfortunately, the kind that interested people who could pay large sums for his offspring. Humphrey had no performance records; he was just himself, an individual who was appreciated by real horse people, but ignored by the big-money horse dealers.

Neither the Cooksleys nor Judith could hire a full-time professional trainer, but Margit, who could compete in the cutting horse market, felt she should show Americans that the Hungarian was a versatile horse. Her husband had died so Margit had to make this decision as well as others alone. She sent Hungarian Miklos and Hungarian Alsab to Ronnie Searles in Arizona. Alsab, a dappled gray, was one of the first of Humphrey's colts. Searles, a fine trainer, won over twenty-five Pleasure and Trail Horse classes with the horses, while Alsab won the Open Cutting in the Southern Arizona Cutting Horse Association.

At the same time, Judith's horses were winning at various shows. Countess Magyar, a bay mare, was Champion Green Hunter in Halifax, Nova Scotia; and other times was Champion Working Hunter and Champion Conformation Hunter.

But the horses received the most extensive publicity under the tutelage of a California trainer, a young woman who was an excellent rider and who knew how to promote horses. Margit heard of her when she won an

Endurance Ride in Oklahoma and asked her to train Hungarians at her Pacific Coast Research Farm in California, which took pupils who wanted to learn training and stable management, as well as riding. The trainer wrote an advertising brochure and, with her husband's cooperation, published articles in "Western Horseman," a popular magazine.

The brochure had a picture of Hungarian Pallo taking a water jump with the trainer in the saddle. Pallo won the training championship in his first Three Day Event, numerous Hunter championships and a Dressage event. The trainer also won the Ladies' Steeplechase at Pebble Beach, riding Hungarian Duma. Hungarian Nicsak, a son of H. Kallo (Bill) and a Hungarian mare, Furioso II, won three Green Hunter Under Saddle classes at Pebble Beach, Ram Tap near Fresno and the Concord Hunter trials. Another of Margit's stallions, Hungarian Brado, won Open Jumping Gamblers Stakes in the Oakland International Horse Show and Champion Hunter at the Concord, California, Hunter and Jumper trials. In addition, Pallo was Middleweight Champion in the Virginia One Hundred Mile Enduance Run, while Brado placed in the Top Ten in the Tevis One Hundred Mile One Day Ride. Many of these horses were sons or daughters or grandchildren of Honpolgar, except Brado, who had been raised by Judith and was by a Thoroughbred stallion out of a Hungarian Halfbred mare.

Thus, Hungarian horses proved they were among the most versatile of breeds, able to compete in Jumping, Three Day Events, Cutting, Dressage and Long Distance Rides, but the Hungarians were still not as well-known as other breeds.

At intervals, Margit told me she had buyers coming to Maryland and to Hamilton to look at the horses, which showed that, gradually, the Hungarians were

Hungarian horse jumping at Pebble Beach, California. Ron Iudice

acquiring a place in the horse world. This did not mean the Hungarians became a household word such as the Arab and Quarter Horses but, after all, it had taken them years to be established and Margit and Judith were comparatively new to showing and selling their horses in America.

When Margit was at the ranch, she was up at dawn and often not in bed until after midnight if she sat with a foaling mare. Judith worked as hard. In addition to a chronic shortage of funds, she sustained an added blow in the death of her husband.

She managed to survive somehow, swamping out box stalls and feeding her horses too much grain, despite the fact that she was in her seventies and didn't move as easily as she had in the past and her eyesight was not what it had been.

One day Margit and I discussed publicity when we rode into the hills to check on the two- and three-year-olds that Margit turned out every summer. This area of the ranch was known as Gird Creek and the pasture the Mountain House. Gird Creek was laced with trails and grass-grown logging roads once used by the Bitterroot Trail Ride so that in a short while we climbed to Mountain House, where we drew rein to watch the colts, pretending fright at our appearance, flee through the sagebrush down into a rocky draw and up to a grassy meadow shaded by huge yellow pines. Brown and bay and dappled gray, they raced with pounding hooves and flying manes. No horseman could have failed to be stirred by the sight.

"Up here," Margit said, "the colts become footsure and develop their muscles and learn to play. A horse that can play is a happy horse and amenable to discipline. In Hungary we believed in letting horses run as much as possible."

By this time there were as many as ten stallions on

the Stock Farm and twenty or thirty mares. Margit selected which stallions were to be bred to which mares, her decisions based on bloodlines, but there was another factor that had not been important before and that was now paramount; that was height. Margit recognized it reluctantly; Judith had been aware of it for some time. Hunters and jumpers had to be sixteen hands or over. The old Hungarian of fifteen two or three could not compete in the market where Grand Prix jumping for large prizes and Three Day Events and Dressage were gaining popularity. Fortunately, Margit had Big John, lent by the Cooksleys, who was by Humphrey out of a Formas mare. Big John was a hunter type and had Humphrey's good disposition. Margit bred him to several Thoroughbred mares that produced colts which grew to sixteen hands and sixteen two. The colts were handsome, but many of them resembled Thoroughbreds more than the traditional Hungarians that had Shagya bloodlines in their ancestry.

There was, however, one event in which the classical Hungarian could compete. That was driving. The Hungarians hold the World Championship, which they have won by entering European competitions.

Their horses are medium-sized, of a Warmblood breeding. The most popular team is the four-in-hand. The horses have natural gaits, free and swinging; they are lively, attentive and muscular. Their manes and tails flow freely. The Hungarian carriage is light, the front wheels smaller than the rear, enabling the vehicle to turn quickly. The coachman sits on the box, the owner or passenger beside him, and the groom or grooms behind. The bits are snaffles. The harness, of leather, is finely crafted with round reins and has metal ornaments, as well as rosettes, fringe and tassles.

Driving competitions, which are held widely in Europe, demand skill and years of practice. To display

Arabian mares drawing carriage at Babolna. Budapest

her Hungarian turnout, Margit selected a team of four white horses and one of bay horses, all about fifteen three hands. These horses Margit had trained to harness in the East by a Hungarian refugee who had driven in his native land. The horses had to learn to do dressage on a proscribed course and to contend in cross-country marathons, where they encountered obstacles such as trees, streams and small, steep hills. Sometimes carriages turned over or became wedged between trees in a tangle of harness and doubletree.

Margit's Hungarian carriage and team were spectacular and were cheered by audiences in New York State, in Connecticut, in Williamsburg and at the National Horse Show.

A prominent magazine devoted to driving featured an article on Margit's Hungarians. Margit sold two of the white horses from one team, but no buyer could afford a four-in-hand and a Hungarian coachman.

Driving was another event in which Margit felt the Hungarians could compete in the eastern market. Shortly after Williamsburg she went to Europe, where she attended a driving exhibition in Hungary. When she returned, she called to say she would be in Montana by the end of the week, but she never came West. She died unexpectedly in New York.

A year later Judith died, also. She was buried on her farm in Virginia where her husband, Dollar, and her mother had also been buried.

Margit was buried in Montana on a hill in the little cemetery overlooking the Stock Farm. It rained during the ceremony when a friend placed a handful of Hungarian soil in the casket before it was lowered into the ground.

Once again the Hungarian horses seemed threatened with extinction. Without Judith's dedication and Margit's hard work and effort there appeared no way the

Margit in Hungarian costume holding Brownie. Ernst Peterson

Hungarians could prosper in America. Rumor had it that the horses would be sold at auction as Marcus Daly's had been and for a while the situation looked bleak. The stud farms had been a business for Margit and Judith. They had been the motivating force behind the breeding, showing and sale of the horses. What would happen now?

But had they failed? Hadn't those two dedicated women left a legacy that would be preserved?

In time, the Cooksleys were able to establish a successful breeding program and were anxious to keep the Hungarian Horse Association an active group. A woman in Wisconsin established a Hungarian Stud Farm. A California couple in Montana bought a stallion and mares for the same purpose. A young woman in Iowa, a talented rider, showed Hungarians successfully. Individuals who could afford only one or two horses bought and listed them with the Association, which had a striking new Certificate of Registration with a picture of the Crown of St. Stephen in one corner and a gold badge ornamented with red, white and green ribbons, Hungary's National Colors. The foreman at the Stock Farm told me he had recently sold horses to Wyoming, Seattle and Ohio buyers for Margit's estate. My daughter, who intends to show in Dressage, bought a colt, a son of Big John and grandson of Humphrey. The colt is a big chestnut and he has a cat he plays with as Humphrey played with his cat in the Edwards' barnyard.

Who can say that Margit and Judith did not accomplish their goal and that, due to them, the Hungarians exist as a breed in America, with the help of a new group of dedicated horsemen? In the future, Hungarians will compete in Grand Prix jumping, in Three Day Events and in all aspects of equine sport except racing. If Judith were here, she would not be surprised, for two

Clovers, granddaughters of old Mokus, graze in a pasture at Bitterroot Stock Farm. They have always survived, for they are proud and have gallant hearts.

Truly, they are the Heavenly Horses.

Appendix

Ancestry of Honpolgar IV (Humphrey)

DUNCAN GRAY

HONPOLGAR (TB)

SHORT HOLIDAY

Imported
HONPOLGAR IV
Gr. 1943
Halfbred

FURIOSO VII

182 FURIOSO VII

254 IBOLYA

POMMERN	POLYMELUS	CYLLEEN
		MAID MARIAN
	MERRY AGNES	ST. HILAIRE
		AGNES COURT
SIBYL GREY	ROI HERODE	LE SAMARITAIN
		ROXELENE
	ROSEY DONOVAN	DONOVAN
		ST. ROSA
HOLIDAY HOUSE	QUEEN'S BIRTHDAY	HAGIOSCOPE
		MATILDA
	HELEN MARY	ST. SERF
		HELEN MAR
SANTIANA	SANTRY	GALLINULE
		E.P.
	GLORIANA	LADAS
		AVILION

FURIOSO XXVIII	FURIOSO XXVI	FURIOSO XXI
		NORTH STAR III
	202 NORTH STAR VI	NORTH STAR VI
		126 FURIOSO XXII
242 FURIOSO XXVII	FURIOSO XXVII	FURIOSO XXIII
		191 NORTH STAR IV
	237 NORTH STAR V	NORTH STAR V
		130 ROUTIER
BOGDANY	WAISENKNABE	BUCCANEER
		SWEET KATIE
	MESSALINA	GUNNERSBURY
		MANELLA
IBOLYA	WAR HORN	BEAUMINET
		WARFARE
	BABONA	237 NORTH STAR III
		BALVANY

THREE AMERICAN BLOODLINES OF HUNGARIAN HORSES

by

Judith B. Gyurky

What to look for in Babolna bloodlines.

Size: 14-2 to 15-3 hands. Neck, elegant and graceful; Oriental type head; good shoulder, short back, good coupling; medium length and harmoniously sloping croup; elbows clear of the body; legs clean and straight, a little in at the hocks is acceptable; medium length withers, and hard, healthy, medium sized hoofs.

Ideal: If 160 cm (15-2) high, 180 cm (71 in.) girth, 19½ cm (7½ in.)

Measurements: cannon bone.

Gaits: Straight, alert, flat, well balanced.

Use: Western work, cutting, endurance, western pleasure, reining, combined training, conformation, hack.

What to look for in the Kisber bloodlines

Size: From 15 to 16 hands and over. Good head and neck, varying according to bloodlines. More length in body than the Babolna lines; very good coupling; a long, or slightly sloping, powerful croup; elbows clear of the body; long, sloping shoulder; good long withers; straight legs, good hocks, and hoofs.

Gaits: Straight, long, elastic, springy, flat "daisy cutting."

Ideal: If 170 cm (16-2) high, 190 cm (74½ in.) gith, 20 cm (7½ in.).

Measurements: cannon.

Use: Hunter, jumper, show hunter, conformation, dressage, endurance, three-day cross country, pleasure, hack, Western.

What to look for in Ujszalla's Gyurky Blood Lines

Size: 15-2 to 16-2, and over. Neck medium; a clean head of good size; good shoulder; long and very strong back; good coupling; long or medium long, slightly sloping croup; elbows clear of body; good bone; very low and long hocks, short cannon, good hoofs; medium, at time low, withers.

Gaits: Easy, well-balanced, and long strided.

Ideal: 170 cm (16-2) high, 190 cm (74½ in.) girth, 20 cm (7½ in.)

Measurements: cannon.

Use: Jumper, field hunter, show hunter, three-day, cross country, dressage.

The Copper King, owned by Bitterroot Springs Ranch, Stevensville.

Judging Procedure and What to Look For

FRONT VIEW

HEAD

Ideal Type: Head well-proportioned to rest of body, refined, clean-cut, with chiseled appearance; broad, full forehead with great width between the eyes; jaw broad and strongly muscled; ears medium-sized, well-carried and attractive.

Faults: Plain headed.

SEX CHARACTER

Ideal Type: Refinement and femininity in the brood mare; boldness and masculinity in the stallion.

Faults: Mares lacking femininity; stallions lacking masculinity.

CHEST CAPACITY

Ideal Type: A deep, wide chest.
Faults: A narrow chest.

SET TO THE FRONT LEGS

Ideal Type: Straight, true, and squarely set.
Faults: Crooked front legs.

REAR VIEW

WIDTH OF CROUP AND THROUGH REAR QUARTERS

Ideal Type: Wide and muscular over the croup and through the rear quarters.
Faults: Lacking width over the croup and muscling through the rear quarters.

SET TO THE HIND LEGS

Ideal Type: Straight, true, and squarely set.
Faults: Crooked hind legs.

SIDE VIEW

STYLE AND BEAUTY

Ideal Type: High carriage of head, active ears, alert disposition and beauty of conformation.
Faults: Lacking style and beauty.

BALANCE AND SYMMETRY

Ideal Type: All parts well-developed and nicely blended together.
Faults: Lacking in balance and symmetry.

NECK

Ideal Type: Fairly long neck; carried high; clean cut about the throat latch; with head well-set.
Faults: A short, thick neck; ewe-necked.

SHOULDERS

Ideal Type: Sloping shoulders (about 45 degree angle).
Faults: Straight in the shoulders.

TOPLINE

Ideal Type: A short, strong back and loin, with a long, nicely turned and heavily muscled croup, and a high, well-set tail; withers clearly defined and of the same height as the high point of the croup.
Faults: Sway-backed; steep croup.

COUPLING

Ideal Type: A short coupling as denoted by the last rib being close to the hip.
Faults: Long in the coupling.

MIDDLE

Ideal Type: Ample middle due to long, well-sprung ribs.
Faults: Lacking middle.

REAR FLANK

Ideal Type: Well let-down in the rear flank.
Faults: High cut rear flank or "wasp waisted."

ARM, FOREARM, and GASKIN

Ideal Type: Well-muscled arm, forearm, and gaskin.
Faults: Light-muscled arm, forearm, and gaskin.

LEGS, FEET and PASTERNS
Ideal Type: Straight, true and squarely set legs; pasterns sloping about 45 degrees; hoofs large, dense, and wide at the heels.
Faults: Crooked legs, straight pasterns, hoofs small, contracted at the heels, and shelly.

QUALITY
Ideal Type: Plenty of quality, as denoted by clean, flat bone, well-defined joints and tendons, refined head and ears, and fine skin and hair.
Faults: Lacking quality.

BREED TYPE
(size, color, shape of body and head, and action)
Ideal Type: Showing plenty of breed type.
Faults: Lacking breed type.

SOUNDNESS
Ideal Type: Sound. Free from defects in conformation that may predispose unsoundness.
Faults: Unsound, blemished (wire cuts, capped hocks, etc.).

ACTION

AT THE WALK
Ideal Type: Easy, prompt, balanced; a long step with each foot carried forward in a straight line; feet lifted clear of the ground.
Faults: A short step, with feet not lifted clear of the ground.

AT THE TROT
Ideal Type: Rapid, straight, elastic trot, with the joints well flexed.
Faults: Winging, forging, interfering, and excessively high.

AT THE CANTER
Ideal Type: Free and easy, and readily executed on either lead.
Faults: Forced and rough.

Hungarian Horse Association
Certificate of Registration

Name/Number ____

Sex ____________________ *Color* ____________________

Markings ____

Sire ____________________ *Dam* ____________________

____________________ ____________________

Foaling Date/Location ____

Breeder ____________________ *Address* ____________________

____________________ ____________________

Owner ____________________ *Address* ____________________

____________________ ____________________

This is to certify that the above named and described horse is accepted and registered with the Hungarian Horse Association

____

Association Registrar

Articles, Unpublished Material, Correspondence

Parade of Horses Recently Imported from Europe. The U.S. Army Remount Service, Front Royal, Virginia, 1946.

Stud Books. Bitterroot Stock Form, Hamilton, Montana, 1965-1985.

Breeders Manual. Hungarian Horse Association, 1966.

Origin and Development of the Nonius. Major Janos Deveni-Mezohegyes, Hungary.

- Broadcast, copy of Radio Free Europe to Hungary, New York, 1954.
- Correspondence Jim Edwards, Army Remount Service, Fort Robinson, Nebraska.
- Catalogue of Horses, Property of Agriculture Remount Service to be sold at public auction, Fort Robinson, Nebraska.

XII Corps. Spearhead of Patton's Third Army. Lt. Col. George Dyer, 1947, History Association.

Before the Colors Fade. Fred Ayer, 1964, Boston, Massachusetts.

The Last Days of Patton. Ladislas Farago-McGraw, Hill Company, New York, 1981.

The Ghosts of Patton's Third Army and *The Houston Break Through.* Bob McCaleb, 42nd Sgdn. A History of the Second U.S. Cavalry.

The XX Corps. It's History and Service in World War II. Prepared by XX Corps Personnel.

The Lippizaners. H.H. Douglas, Armored Cavalry Journal, 1947, Washington, D.C.

Horses and Mules and National Defense. Department of the Army, Office of the Quartermaster General, 1958, Washington, D.C.

Hungarian Horses in America. Tex Johnson, 1974, Unpublished article.

Books

The Horse of the Desert. Compiled by Major General James Harbond, 1967. Jay Schuler Company, Springsfield, New York.

The Thoroughbred. Peter Willetti, 1970. Weidenfield and Nicholson, 5 Winsley Street, London.

The History of Horsemanship. Charles Chenevix-Trench, 1970. Doubleday and Company, Garden City, New York.

The Arabian Horse. Edited by George Conn, 1959. Arco Publishing, New York.

What Horse for the Cavalry. Spencer Borden, 1912. J.H. Franklin Company, Fall River, Massachusetts.

Magyar Los Portes Lotenyesztes. Bossanyi-Havas Rezso, 1943. Budapest, Hungary.

The History of Horse Racing. Roger Longrigg, 1972. Stein and Day, New York.

A History of Hungary. Dominic G. Kosary. Benjamin Franklin Bibliophile Society, 1941. Cleveland, Ohio.

They Rode Into Europe. Miklos Jankovich, 1971. George Harrap and Company, High Holborn, London.

Breaking and Riding. James Fillis Blackett, Ltd. Paterngster House: London.

The Mark of Clover. Judith Gyurky, 1953. Jarmin Printing Company, Charlottesville, Virginia.

Distance Riding. Virginia Johnson, Thula Johnson, 1976. Houghton-Mifflin, Boston, Massachusetts.

My Horses, My Teachers. Alois Podhajsky, 1968. Doubleday and Company, New York.

Ungarns Pferde. Morique and Hans Dosenbach, Ivan Hatos. Hallwag Verlag, Stuttgart, 1976.